# The Politics of Place Naming

SCIENCES

*Geography and Demography*, Field Director – Denise Pumain

*Political Geography*, Subject Head – Frédéric Giraut

# The Politics of Place Naming

## *Naming the World*

*Coordinated by*

Frédéric Giraut

Myriam Houssay-Holzschuch

iSTE

WILEY

First published 2022 in Great Britain and the United States by ISTE Ltd and John Wiley & Sons, Inc.



ISTE Ltd
27-37 St George's Road
London SW19 4EU
UK

www.iste.co.uk

John Wiley & Sons, Inc.
111 River Street
Hoboken, NJ 07030
USA

www.wiley.com



Library of Congress Control Number: 2022943330

British Library Cataloguing-in-Publication Data
A CIP record for this book is available from the British Library
ISBN 978-1-78945-115-3

ERC code:
SH2 Institutions, Values, Environment and Space
*SH2_11 Human, economic and social geography*

# Contents

# Acknowledgments

**Frédéric GIRAUT[1] and Myriam HOUSSAY-HOLZSCHUCH[2]**

[1] *UNESCO Chair in Inclusive Toponymy "Naming the World", University of Geneva, Switzerland*

[2] *Université Grenoble Alpes, CNRS, Sciences Po Grenoble, Pacte, France*

The editors would like to thank UNESCO for its support in developing their activities and international networks in the field of this book, thanks to the status of UNESCO Chair granted to the Chair in Inclusive Toponymy "Naming the World" based at the University of Geneva. The Chair is preparing an original massive open online course (MOOC) in parallel with this book. We would also like to thank the Department of Geography of the University of Geneva for financing the styling of the chapters before submission to the editor and Olivier Lavoisy for the creation of this initial styling. Finally, we would like to thank all the authors for their valuable expertise.

# 1

# Naming the World: Place-Naming Practices and Issues in Neotoponymy

**Frédéric GIRAUT[1] and Myriam HOUSSAY-HOLZSCHUCH[2]**

[1] *UNESCO Chair in Inclusive Toponymy "Naming the World", University of Geneva, Switzerland*

[2] *Université Grenoble Alpes, CNRS, Sciences Po Grenoble, Pacte, France*

Naming the world is very much the adventure societies, groups, individuals and humans engage in when they inhabit a portion of earth, a territory, a city, or they follow routes, discover them, invest in them, or map them. This takes very different forms according to whether someone is a hunter-gatherer, a farmer, a navigator, a merchant, a conqueror, an engineer, or an administrator. Depending on whether someone is colonizing a frontier or claiming back ancestral lands, whether or not they have a written form of their language, whether they are part of a collective with shamanic approaches or work for a modern state, or whether they plan a serviced space or produce and inhabit an informal space, the naming operation is never definitive nor exclusive; it is in any case fundamental.

Are the Aragonese peaks the world of their so-called Pyreneist discoverers, somewhat settlers and cartographers of the peaks, whose names stuck to these eminences in the 20th century, or are they the world of the peasant populations of the valleys and high mountain pastures that used them as a starting point for religious and cosmogonic meanings, and named the passes and slopes instead? The Aragonese government has resolutely opted for the second option, renaming all the peaks above 3,000 meters after Aragonese names borrowed from the local and

*The Politics of Place Naming*,
coordinated by Frédéric GIRAUT and Myriam HOUSSAY-HOLZSCHUCH. 

peasant microtoponymy. The operation, a toponymic restoration, was carried out by transferring or extending vernacular names of practiced places (passes and slopes) to the peaks thus seen as "indigenized" and "decolonized". Another option, probably that of the majority of users-inhabitants and hikers, will be to refer alternately to each of the naming systems, making these worlds of meaning cohabit at the cost of a certain practical confusion. For naming participates simultaneously in the practical and magical organization of situated life, in heritages processes and in individual and collective identities.

## 1.1. Political/critical toponymy: an emerging field at the core of territorialization issues

Thus, toponymy appropriates places and spaces through addresses, landmarks, referents, routes, signage and cartography. In a word, it territorializes. Place names are indeed at the core of the operations of territorialization; it is the soulful bit that transforms a habitat into an inhabited place in the full sense of the word, a space invested with meaning and therefore a territory. Claude Raffestin defined the territory as follows in 1986:

> The territory is a reordering of space whose order is to be sought in the informational systems available to man [sic] as he belongs to a culture. The territory can be considered as space informed by the semiosphere (semiosphere = set of signs); all the mechanisms of translation, which are employed in the relations with the outside, belong to the structure of the semiosphere (Raffestin 1986, p. 177).

The activity of naming and renaming places, *neotoponymy*[1], as well as making names visible through signage and cartography, associated with the linguistic and functional use of place names, in other words, the toponomascape[2], participates fully in this production of territory through the semiotization of space.

Naming thus territorializes, politicizes and ideologizes because it exists in itself as a sphere (that of the toponomascape) that possesses its own logic, its own autonomy, which has as much to do with the question of power, property (Ritchot 1989) and technique as with representation and identity. The meaning of naming is

1. The concept of *neotoponymy* refers to the activity of producing new place names by naming new places or by renaming, but the term also refers to the corpus constituted by these new names.

2. The *toponomascape* is thus made up of all place names, their linguistic use and their figuration in signage, addressing, cartography, nomenclatures, databases, discourses, practices, etc.

not only in what we finally designate; it is also to be sought on the side of what we designate or do not designate, of what makes us designate, of those who designate and of what we do with the designator.

The naming of places is fundamentally political because it participates in the institution of an order endowed with a regime of representation that hierarchizes places, resources, values and beliefs. It is subject to debates, to controversies, to claims, passions and conflicts, but it is also a discrete component of symbolic and practical order that insinuates itself in everyday life and pervades it. It is constitutive of the civil status, of the address – therefore of the identity – and of the system of location, and by that it shapes the individual and collective relationship with the places.

Thus, the study of place naming, the naming of the World, constitutes an autonomous field, distinct and complementary to that of classical toponymy, which is a branch of onomastics or the study of proper names in linguistics. It is an emerging field, referred to as critical toponymy or political toponymy or place naming studies, and this book outlines its contours, issues, contributions and perspectives.

While classical toponymy studies the composition, meaning and alteration of place names by mobilizing etymological, philological and pragmatic approaches, political toponymy studies the social, economic and cultural stakes involved in their choice and use. The study of place naming and its issues is also different from toponymy as an indicator of human occupation and environmental changes, allowing for an archeology of settlement and/or environment dynamics.

However, the links between toponymy, settlement and the relationship with the environment also provide information on the contradictory motivations and legacies of naming, and are of direct interest to political toponymy in this sense. This is what **Sébastien Boillat** presents in this book, by pointing out the power issues concerning microtoponyms in different environments, and how much they reveal about the environmental perception and ethics of the inhabitant groups, particularly the Indigenous populations.

The importance of toponymic production is, however, often disputed. However, each time the question of the name is a public issue, it unleashes passions and reveals potentially contradictory positions and representations. The debates, controversies and counter-initiatives then draw a geopolitics of variable scale. But the relevance and relative importance of toponymic production among the social issues and political choices considered as priorities are also systematically questioned, contested and devalued. This questioning and the relativistic and anti-neotoponymic discourses comprises elements of the reactionary rhetoric dissected by Albert Hirschman (1991): the inanity of the subject (in other words, its anecdotal nature or futility); its adverse effects in terms of costs and functionality, and the threat to social, political and cultural equilibrium. This systematic contestation of the naming question in the public debate, in spite of the mobilizations around it, has made the affirmation of the related scientific field more difficult.

## 1.2. Political toponymy: a recent history?

The history of critical or political toponymy – which we do here from a corpus unfortunately limited for linguistic reasons to the literature in English and French – is indeed recent. While reflections and observations on the link between power and place names may have accompanied works of classical toponymy or descriptive geography, such as accounts of conquest like Julius Caesar's *Gallic Wars* or those of the administration of an empire like *Historia Augusta* (included in the literary biography *Memoirs of Hadrian* by Marguerite Yourcenar in 1951), publications entirely dedicated to the question of the political stakes and techniques of naming are contemporary. In addition to Gramsci's work in *Prison Notebooks* that makes street names elements of hegemonic domination (see infra; Vuolteenaho and Puzey 2018), two works from the first part of the 20th century appear to be true pioneering works before the explosion of studies in the 1980s and the attempts to mark out the field in the 21st century. These works are Lucien Gallois' *Régions naturelles et noms de pays* in 1908 and George R. Stewart's *Names on the Land: A Historical Account of Place-Naming in the United States* in 1945.

In the first work, the author starts from a questioning on the adequacy of the names of French rural regions with knowledge and peasant traditions, academic knowledge and political divisions. Thus, for the Brie region, Lucien Gallois notes:

> For the geographer and the cartographer, the Brie became a sort of territorial subdivision to which it was necessary to give precise limits and it is thus that one understood it between the Seine and the Marne. The farmer has other preoccupations, it does not matter to him that the Brie was originally a forest, what he calls thus is the plateau which crowns the cereal lands; he does not feel the need to delimit it exactly, but he knows successfully how to distinguish the good and the bad parts of it (p. 138).

He comes to believe that the uses and promotion of these names of regions always have to do with questions of representations and stakes to be contextualized. Jean-Claude Chamboredon who proposes an exegesis of the text in 1988 concludes:

> On the basis of spatial variations, of diverse historical-natural units, unequally relevant according to the domains (agricultural economy, landscape, mobility, etc.) and unequally present in common practice and perception, the work of naming specialists (from academics to local scholars) who, if necessary, borrow popular terms, progressively constitutes categories that can come back to confirm, with the authority of a scientific foundation, the vernacular designations. Describing and naming, they define, according to a modality that one can say performative, territorial resources as spaces of adhesion and forms of identity (p. 34)[3].

George Stewart's work was addressed to the general American public after the war, inviting them patriotically to immerse themselves in the ingenuity and diversity of the American melting pot through colonial and pioneer neotoponymy. In doing so, he was already tracing the possibility of a reflection on toponymic and territorial creation and translation in a context of plurality of legacies, representations and actors[4].

It was not until the 1980s that a number of studies of political toponymy were published that examined the stakes and actors involved in naming. The very first known publication of a monographic study in this field dates from 1977. Inge Kleivan invokes cultural imperialism in her analysis of Greenlandic neotoponymy. She shows how, after Norwegian imperialism, Danish imperialism imposed an indigenization of neotoponymy for the benefit of its southern community representatives. The Middle East, Africa and Oceania were then the subject of pioneering publications that opened up different avenues. South Africa, constantly giving rise to studies in the field, introduced with Gordon Pirie's text (1984) the bureaucratic creation of the name of Soweto, a political concern in the deciphering of the neotoponymic factory in a highly segregated, technocratic and racist context. Roland Pourtier (1983) was interested in the creation of a postcolonial state toponomascape by identifying continuities and divisions in the naming of regions and districts in French-speaking Central Africa, particularly in Gabon, where the newly independent state systematized a naming policy based on a "neutral" hydrographic reference system in contrast to cultural groups and their territorial anchors. The Middle East, with the two neotoponymic hotspots of Iran (Lewis 1982)

3. The geographer Alain Reynaud and the linguist Philippe Gardy took up such an approach, respectively, for Larzac (1987) and for Champagne (1990), which has become the place and name of the dispute.

4. Wilbur Zelinsky (1967, 1983) and Christian Montes (2007) continue this line of analysis of the production of the American toponomascape by Euro-Americans.

and Israel/Occupied Palestinian Territories (Cohen and Kliot 1981, 1992), also appears in works on toponymic nationalism and its geopolitical declinations. Starting with Oceania, Paul Carter (1987) introduces a complementary reflection on colonial toponymy and its motives (see the chapter on conquest toponymy in this book). Also, the problem of indigenous toponymy is addressed through contemporary power issues within customary authorities in Vanuatu (Rodman and Rodman 1985). In Canada, as academic knowledge of indigenous geographical knowledge progresses (Collignon 1996), the question of land claims and associated toponymic knowledge and maps becomes an object of study (Wonders 1987; Müller-Wille 1989).

The question of the invisible toponymies of subaltern populations was thus launched, and it was to be the subject of spectacular research during the 1990s and 2000s (Stickler 1990; Nash 1993; Berg and Kearns 1996), particularly in settler colonial contexts, and especially in South Africa, where the already burning issue became crucial with the end of apartheid (Jenkins 1990, 2007; Giraut et al. 2008). The invisibility of toponymy is a major political issue today with the question of toponymic restoration and revision in a perspective of inclusiveness. Thus, the question of colonial names that negate or recuperate indigenous names is the subject of much work (see Giraut's chapter), which is not yet the case for the question of gender in commemorative and more generally in official toponymy. At the same time, there has been an opening towards private actors in real estate development and the imaginary associated with the world of suburban housing or tourist resorts (Debarbieux and Gumuchian 1987; Perkins 1989; Hopkins 1998; Wood 2002). These works can be considered as precursors of those that would come to deal with the toponymic commodification and the economic value of place names, as well as with metropolitan storytelling and, more generally, territorial marketing.

Since the mid-2000s, major advances have been made towards the constitution of a real field around original and potentially interdisciplinary themes. Above all, the field is becoming aware of itself and is developing an agenda and theoretical concerns. The corpus of studies is considerably enriched with collective publications in the form of special issues of mainly francophone journals (*Espace géographique* 2008; *Espace politique* 2008; *Geojournal* 2008; *ACME* 2011; *Droit et culture* 2012; *Echogeo* 2020) and edited books (Bouvier and Guillon 2001; Guillorel 2008; Berg and Vuolteenaho 2009; Bigon 2016; Puzey and Kostanski 2016; Rose-Redwood et al. 2018; Caïazo and Nick 2020) that compile a number of case studies and propose transversal readings.

It is also worth noting, at the beginning of the period, the publication of textbooks (Randall 2001; Kadmon 2000; Monmonier 2006) which approach the question of toponymy from the angle of geopolitical issues in order to reaffirm and illustrate that the imposition or claim of geographical names is a question of power and imperialism by proposing an inventory of different practices (Houssay-Holzschuch 2008).

Among the original themes that were asserted during the period, most of which are summarized in this book, one of the most represented is the question of commemorations via odonymy[5]. The context of the end of the Cold War, with a number of radical changes in political regimes, particularly in central and eastern Europe, led to massive changes in place names and particularly in the names of urban streets. These practices could be interpreted as state interventions aimed at imposing new political referents and above all at reversing an inherited and reviled frame of reference for landmarks. The work of Almeida-Topor (1996), Azaryahu (1996), Yeoh (1996), Light et al. (2002), Bigon (2008), Palonen (2008), Bourillon (2012) and Njoh (2017) has been particularly salient in the study and interpretation of these practices in different contexts. Following Milo's seminal work on odonymy as places of memory (Milo in Nora 1984), Azaryahu has given a comprehensive interpretation of these processes (2011) after **Alderman** (2008) integrated them into a broader reflection on issues of heritage and memorialization. In particular, in his chapter dedicated to toponymic commemoration, Alderman works on the meaning of the challenges to past commemorations related to colonialism and racism, based in particular on the US experience. Alderman lists the major questions related to this issue: "What political actors, decisions, and social relations enable certain memorialized names to be displayed publicly? Who is given the power and authority (or not) to do this work of remembering and naming? Whose identifications with the past are enacted (or denied) through place naming? How do these toponyms work to benefit (or disadvantage) the belonging and social standing of certain public groups over others?"

These questions point to many of the programmatic elements proposed in the seminal article by Rose-Redwood, Alderman and Azaryahu (2010), which was followed by a special issue of the journal *ACME* (2011). In these publications, the authors identify a scientific agenda for critical toponymy, in addition to methodological reflections and proposals concerning actors, scales (Hagen 2011) and sources. This agenda proposes three main, non-exclusive paths: "political semiotics, governmentality studies, and normative theories of social justice and symbolic resistance" (Rose-Redwood et al. 2010), paths to which the following section returns.

---

5. It was itself methodologically and historically marked out in the French context by Badariotti (2002) and Bouvier (2007).

## 1.3. On the agenda of political/critical toponymy: contradictory promotion of functional, market and inclusive corpuses

A political semiotic approach points to the well-demarcated field of political odonymy. This field has been enriched by works on the naming and visualization of new emerging territories, resulting from the politico-functional recompositions of new regionalism (Giraut and Houssay-Holzschuch 2008b), or of tourism territories (see **Gauchon**'s chapter) or transport (**Destrem**'s chapter) as places where the political hold of the visual and linguistic landscape extends, together with territorial marketing in a globalizing context. At the same time, collective publications in socio-linguistics, drawing inspiration from visual studies, geopolitics and semiotics, address the linguistic landscape and toponymic issues, in particular in plurilingual situations (Gorter et al. 2011; de Vergottini and Piergigli 2011; Blackwood et al. 2016; Cornips and de Rooij 2018).

This political semiology approach converges with governmentality studies in the Foucauldian sense through the questions of controlling and tracking populations via the identification of places of residence, as well as the commodification of place names (Light and Young 2014; Medway and Barnaby 2014). In this volume, **Reuben Rose-Redwood, Anton Tantner and Sun-Bae Kim**'s chapter offers a long-term perspective on addressing practices that includes numbering and openings to digitization. Jani Vulteenaho, in the chapter devoted to the commodification of place names, proposes a conceptualization of the relationship between ownership and place names that allows us to differentiate practices, issues and toponymic implications.

The third approach, that of social justice issues, in other words, of toponymic inclusiveness, is particularly topical with the overlapping decolonial and feminist demands. These revisit the question of toponymic rights and wrongs and argue for a more equitable toponymy, redressing deliberate and abusive erasures – those of Indigenous people, racialized communities, women and queer people, and more generally of the sabalterns in patriarchal and postcolonial societies. Inventories of recognition and visibility deficits and their historicization are becoming more numerous (Berg 2011; Novas Ferradas 2018; Bigon and Zuvalinyenga 2021; Beck 2021; Zuvalinyenga and Bigon 2021).

Beyond the toponymic injustices related to race and gender (that of class being well known since Gramsci), questioning dominations in naming opens up an important field of research concerning vernacular, alternative and informal toponymies, which could take over from a certain focus on official toponymy, its state actors and its referentials. Thus, the naming of places of so-called spontaneous or informal settlements in cities of the Global South is becoming an important topic (see Melissa **Wanjiru**'s chapter in this book), be it indirectly in an attempt to understand resistance to addressing (Njoh 2010; Ndock Ndock 2020), or directly for its own sake and its theoretical potential, based almost exclusively on African experiences in the Francophone and Anglophone literature (Bertrand 2001; Dorrier-Apprill and Van den Avenne 2002; Leimdorfer et al. 2002; Boyer 2014, 2017; Cumbe 2016; Wanjiru and Matsubara 2017; Mokwena 2020; Irvine et al. 2021). In parallel, and especially in Europe, anthropological work is devoted to popular practices of counter-naming or uses of alternative toponyms, be they slang or internal to a group and/or counter-culture (Chaté 2005; Steffens 2007; Lajarge and Moses 2008; Paunonen et al. 2009; Veschambre 2011; Vuolteenaho et al. 2019; Bing 2020; Vivant 2020; Wright 2020).

Other studies examine the activist and proactive dimension of inclusive toponymic claims as a social movement that questions the nature of political mobilizations through the double lens of intersectionality and performance studies (Rose-Redwood 2016; Brasher et al. 2017; Beaudouin and Martin 2019; Xu 2021).

With the end of sovereignty over cartographic geographic information, the cartographic scene, and more generally the geoweb, now appears as a potential field for the promotion of alternative toponymic corpuses. But studies going beyond a state of the art and the causes are still rare (Chatelier 2007; Choplin and Lozivit 2019; Casagranda 2020; Ndemo 2020; Palmer and Corson 2020; Carrie et al. 2021). Mathieu **Noucher** offers a chapter on this issue in this book and illustrates it with an enlightening case where actors involved in a public program to collect indigenous microtoponyms in French Guiana must eventually resort to the OpenStreetMap contributory mapping platform to integrate it into the visible toponomascape.

The obvious links between these three approaches are woven together in this book and should belong to the agenda of critical toponymy: the multiple or the simultaneous, plural designations of places, the relational or the overcoming of the exo/endonymy binary, and the questioning, or rather the overcoming by participation of the official, state, paths of naming and mapping. For example, and with regard to the first two points, Bakary Traore (2007) has shown for the West of Burkina Faso, which actors have historically designed a double toponymy, and how: commercial and relational naming in the cities; locally anchored and descriptive of a sacred and lay socio-environmental order in the countryside.

More broadly, one extraordinary case alone illustrates the relevance of these interdisciplinary dimensions: place naming after Nelson Mandela. Today, Nelson Mandela is probably the most widely used place name in the world (nearly 3,000 occurrences recorded as names of publicly accessible places, if we add his clan name Madiba). Does it concern the whole world? No, mainly the Black Atlantic, to follow Paul Gilroy's expression, illustrating the strong political and cultural links between communities and places marked by the history of colonization, the Atlantic slave trade and Black cultures. The toponym was first used in the 1980s, in the midst of the Cold War, evoking a then imprisoned leader, a hero in the fight against the late, spectacular and terrible figure of colonization that was apartheid. This first diffusion happened in very committed and concerned European or American places of contestation and counter-culture: campuses, red belt suburbs or Black ghettos, already creating links of commitment between heterogeneous and distant places. An exponential diffusion then takes place in two directions. The first constitutes a demand for and by the marginalized or relegated, mainly but not exclusively Afro-descendants, who name (spontaneously or as a claim) their camps, informal neighborhoods, collective housing projects or places and community facilities after him. The other direction, composite and very broad, points to a certain political or economic commodification. A Nobel Peace Prize winner and an icon of dialogue between populations and of good governance, the largely ecumenical Mandela figure becomes a kind of brand that allows for the joint celebration of the universal values of human rights and those of a happy globalization that includes Africa. Such a symbolic load makes the name into a resource and a powerful vector of territorial, political and commercial marketing. Since global cities, as well as secondary cities aiming at putting themselves on the map, add references in the public and political nomenclature of roads and facilities as well as in the landscape of private brands. Nelson Mandela is everywhere in the toponomascape, simultaneously embodying the aspirations of progressive and conservative elites and those of the outcasts of the outskirts, within the same metropolitan areas, from the center to the peripheries. Place naming might be bottom-up – a collective called upon by participatory processes to name a new school, a movement of landless people placing themselves under Mandela's patronage to occupy a vacant lot, an informal entrepreneur opening a motorcycle cab station or a registered private management school. But it is also often top-down and the result of a political decision, even if this does not guarantee a successful take: in Douala, the users of the roundabout "*J'ai raté ma vie*" (I messed up my life) do not intend to let go of this vernacular name that ironically shouts to the world the tragedy of the poor and marginalized lives of the neighborhood in exchange for a parachuted Nelson Mandela toponym. On the other hand, the initiators of clandestine funk dances in Brazilian favelas produce in all working-class neighborhoods across the country ephemeral events and places (the *baile do Mandela*) referring to its origin in the eponymous favela of Rio. A fantastic diffusion of the same referent, both exonym and endonym, which generates a bundle of relationships and oppositions, of simultaneous and contradictory uses, of initiatives and crisscrossing performances, contributing to the variable geometry of globalized metropolitan imaginaries.

This particular and heuristic case mirrors the field of critical or political toponymy: it is indeed proliferating. So, the final question that we must ask ourselves in this introductory overview is surely how to theorize from such a profusion. And this is indeed the most burning issue for this emerging field.

## 1.4. Theory-in-progress: beyond hegemony and *dispositif*, a toponymic situationism?

A first challenge rises in relation to linguistic approaches, as the postulate of the toponym as a "rigid designator" being questioned (Kripke 1972; Recanati 1983). Critical toponymy reveals it to be contingent, descriptive and evolving. It is therefore descriptivist, in that it grants the toponym a set of characteristics of the place it designates, and pragmatic, which grants the place name a meaning and implicatures that are sensitive to the context of use (Laugier 2004; Moeschler 2019). Better still, it increasingly affirms its performative dimension (Austin 1963; Searle 1963): in other terms, the place name has the capacity to produce social effects depending of its use. Shifts in meaning, designation and use, and resemantizations (Kristol 2002) are thus emphasized and worked on in context. Toponyms that refer to banal places might become landmarks, names that evoke and travel, for example, the names of battles (Paveau 2008) and ideological confrontations such as Larzac (Gardy 1987), the precursor of the "*Zones à Défendre*[6]" such as Notre Dame des Landes, or tourist landmarks such as Lascaux (Neotopo 2020), or of contemporary popular songs such as "Penny Lane" (Gensane 2010; Neotopo 2014). These may be references to landmarks or popular figures whose metonymization is reinforced by associating several toponyms (Charbonneaux 2020). The case of Nelson Mandela is again exemplary, as his name is often caught up in toponymic associations that situate him within a specific pantheon (together with Martin Luther King, Mother Teresa and Gandhi, more generally with the leaders of decolonization and/or the civil rights struggle, or Nobel Peace Prize winners), or as an actor on the South African scene (associated with Frederik de Klerk or, by contrast, with other anti-apartheid fighters, or even with the name of Soweto, which is associated with the struggle). But mundane names and their negotiation are also a prime field for an application of metapragmatics approaches that work on stakeholders' skills of analyzing the issues and effects of language (Karlander 2017).

---

6. Literally meaning "Zone to Defend", it is a French neologism used to refer to a militant occupation that is intended to physically impede a development project. By occupying the land, activists aim to prevent the project from going ahead.

We should also note the recent development of quantitative methods in critical toponymy (Eades 2017; Oto-Peralias 2018; Badariotti et al. 2021; Bancilhon et al. 2021; Fabiszak 2021). They might not self-identify as such, rather as digital humanities, but fully address the social, cultural and political importance of toponymic choices and uses. Even so, these quantitative approaches return to a more fixed and rigid conception of the designator constituted by the toponym, geo-referenced and included as a unique and perennial entity in a nomenclature and/or database. The emphasis is therefore not on the process of naming or renaming but on the unique referents of the place name, within Millian ontologies (Stuart-Mill 1896) and an analytical philosophy of language. A critique of these approaches could focus on the impoverishment of the social value of the toponym when it is reduced to the status of a rigid designator, and on the impossibility of processing toponymic corpuses to extract a global meaning relative to their presence, importance and diffusion, without contextualizing the naming process and accounting for evolving meanings. Finally, the notion of city text (Duncan 1990) allows us to find a common conceptual ground when dealing with odonymy (Palonen 2008; Oto-Perialas 2018; Smith 2018). Some theoretical divergences remain beyond methodological common ground, but allow for diverse contributions, identifying regularities and irregularities that sustain the debate.

Interpreting the origin of the place name, its motivations and modalities, is another issue addressed by linguist toponymists that is of great interest to political toponymists. The recent debate between Tent and Nash (Tent and Blair 2011; Nash 2015; Tent 2015) actually validates what seems to be a consensual, if not hegemonic, result of critical toponymy: the need to take into account the entire naming process and not just the act that establishes the place name. What Nash calls the "how?" must complete and take precedence over the "where?" the "when?" the "who?" or even the "why?" of naming.

Within critical/political toponymy itself, theoretical progress happens incrementally rather than by controversies. Inspirations from varied social sciences allow us to highlight different aspects of the production, function and use of place names.

Sociology is sometimes called upon (Rusu 2020, 2021). The Bourdieusian concepts of field, and especially of capital, are regularly used, even diverted, notably to mark out theoretically the question of the commodification of place names: place names have multiple values, depending on the fields that constitute them as various types of capital, be it for the place owners and/or their users.

Political science is frequently called upon when it comes to understanding the State and the administration investment in toponymy, through self-centered bureaucratic modernist logic (Scott 1998) and the imposition of an everyday, banal nationalism (Billig 1995). Official toponyms are by far the most visible and the most worked upon. More generally, Gramsci and his concept of cultural hegemony as a component of class relations are a frequent and powerful reference to analyze the state toponymic production in terms of intentionality, means and effects (Puzey and Vuolteennaho 2016).

Foucault is also a common and fruitful reference. In particular, the Foucauldian prism of biopolitics is used for the calculation of the world and the intrinsic link between toponymy, addressing and population control through the adoption of practical systems (Rose-Redwood 2011). Toponymic standardization as a means of geolocation in the context of big data offers new possibilities for the development of this theoretical field.

The Foucauldian concept of dispositif is also useful to apprehend the toponymic factory in its various dimensions. Moreover, it can be perfectly combined with the Gramscian notion of hegemony, as a means of its deployment. Our own analytical and interpretative grid (Giraut and Houssay-Holzschuch 2016) uses the notion of dispositif. It is based on extensive bibliographical work on political toponymy monographs, studying various periods of time and situated at very different scales and in different contexts on all continents. Our long-term work on official, community and conflictual nominations in South Africa has also provided fertile ground for a type of theorizing that combines different logics.

First, using the notion of dispositif reaffirms the need to understand the toponym not only per se, but also to systematically situate it within the naming process, which also becomes an object of research. In other words, with the notion of dispositif, the toponym as well as the naming of the place can be interpreted in political terms. The toponymic dispositif thus takes toponyms as objects, nomination as a process and the toponomascape (made up of place names as they are deployed in space or on maps, as well as their referents) as a whole or as a field of governmentality.

In our modelization, the toponymic dispositif combines four types of components: geopolitical contexts; technologies, frameworks and purposes of naming; places, which can be invested in differently according to their status; and the actors who invest in the field in possibly contradictory ways.

We identify four generic contexts that summarize the main historical and geopolitical situations of naming or massive renaming: conquest (imperial, colonial, pioneer, as well as local or regional expansionism); revolution in the sense of the radical change of a political order and its associated referentials (e.g. decolonizations or, more modestly, the advent of new local executives); emergence in the sense of the establishment or affirmation of new localities or new territorial entities when political power is exercised, whether they be cities, new States or local governments, or, on another scale, neighborhoods, housing projects or resorts, etc.; commodification confers a market value that can be negotiated to the toponym.

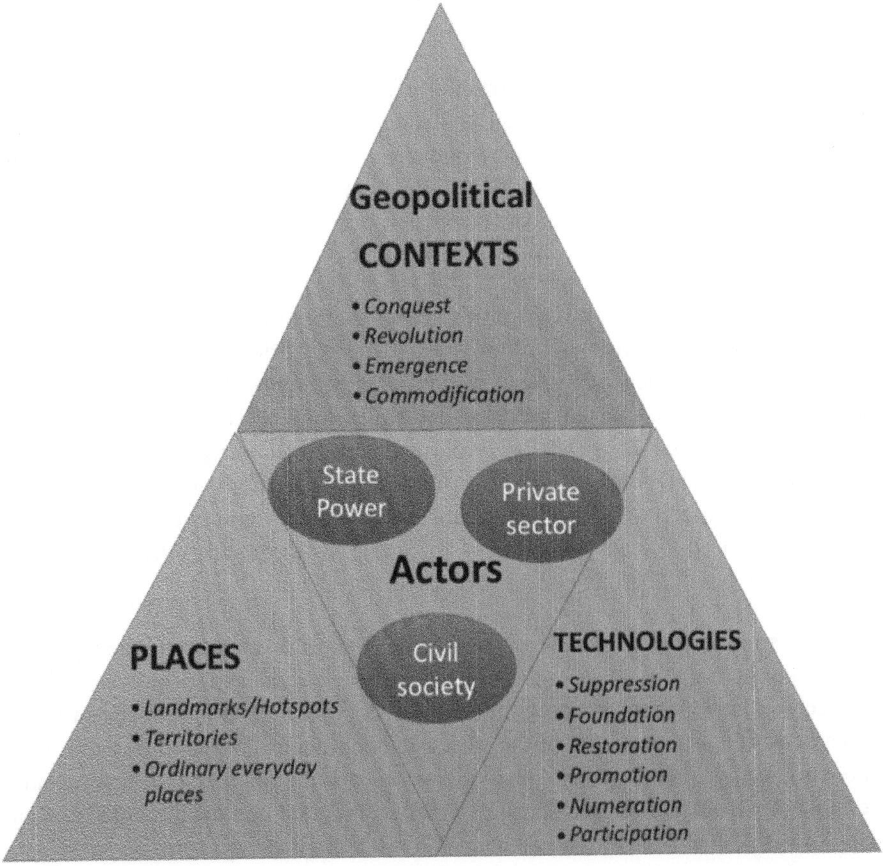


**Figure 1.1.** *A grid for interpreting the political toponymic dispositif according to geopolitical contexts, toponymic technologies, actors and places (source: from Giraut and Houssay-Holzschulch 2016)*

The numerous political technologies of naming associate techniques per se and the objectives of nomination. They are to be understood in the Foucauldian sense of practical rationalities governed by a conscious objective. We have reduced technologies to a few in our model, and associate them with contexts. The technology of suppression, or of toponymic cleansing, aims either at symbolically appropriating conquered places (conquest) or at purging the toponymic landscape of the ideological referential of an old regime (revolution). Conversely, the technology of the toponymic foundation aims to assert a new referential in a context of revolution or conquest, or even emergence or commodification. Toponymic restoration consists of a political undertaking to restore names or references that may have been denied or erased. This often happens in contexts of conquest, revolution or even emergence. Finally, the technology of toponymic promotion, which is particularly connected to the contexts of emergence and commodification, covers the transfer of naming rights of a public place to sponsors for advertisement purposes, or the marketing of a place or a destination. Participation technologies associate different stakeholders in the naming process. They are often an initiative of public actors, target civil society and call on private actors as consultants. Naming roads and places by numbers or letters is a practical technology well adapted to the contexts of emergence and urban creation. It can, however, constitute a marker of marginalized spaces, for example, in segregated contexts such as the townships of apartheid South Africa.

This example shows the importance of a specific place, its nature and status, which differentiates the application of technologies across contexts. Thus, landmarks will be particularly targeted when going for foundation, but banal places, the small local streets, for example, will also carry an ideology or promote a specific referential, even if only by the language they use. Finally, territories (spaces corresponding to a more or less formal jurisdiction) can try to distinguish themselves, affirming their own identity through foundation or promotion.

Of course, these typical connections between the components of the dispositif are central to our theoretical proposition; they are orchestrated by the interplay of actors, who also need to be situated. For modern and contemporary history, we have reduced the main actors to the spheres of the state (comprising bureaucracy and local government), civil society (comprising organized collectives and experts) and the private sector. The latter does not only appear in contemporary settings, though, for example, the naming of sports arenas: private actors were instrumental in naming city roads in the Middle Ages.

Our theoretical and verified grid offers a tool for modeling, interpreting and comparing complex situations and experiences, but it also restores the complexity and processual nature of each situation. This process is deciphered as a singular nexus of contexts and an interplay of actors deploying technologies that are place-dependent.

The main theoretical challenge, as this book will hopefully make clear, is to go beyond simple binaries: endo-exo, top down-bottom up, producer-user. For this, we draw inspiration from the Southern Turn, which challenges social sciences by centering Southern theoretical and epistemological experiences. For us, African experiences are especially crucial: they show the limits of approaches, analyses and interpretations forged in the North, and provide situations and experiences that deny the primacy of the state, the formal and the normalized. Deserting the "author for the reader" (Düzgün 2020, from Barthes), the nomination for the reception, the landmarks for the ordinary everyday places, the formal for the informal (Bigon 2020), the imposition of a name for its use and appropriation (Ben Arrous and Bigon, in this book) as the horizon of critical/political toponymy refers of course to Michel de Certeau and his "invention of the everyday" (1980) and much more generally to "ways of doing" and performing. It centers concrete practices and uses that might include how place names are materially inscribed in visual space and in the concrete spatial practices that mark inhabited places and circulations.

A barely explored reference (Vuolteenaho and Kolamo 2012; Kasi-Tani 2014; Neotoponymy [blog]) that seems to us potentially very fertile is situationism (Debord 1967; Vaneigem 1967). This movement of thought and action is a powerful conceptual framework for critical/political toponymy. On the one hand, it allows for a deeper understanding of the logics that presides over toponymic production. The toponomascape of everyday life can be interpreted as the toponymic projection of the norms of a desirable life. These norms, which form the basis of the contemporary spectacle, are incorporated, claimed or reproduced from below, by the spectators who participate fully in the construction of the scenery. For instance, they participate in the diffusion and appropriation of a pastoral suburban imaginary, inspired by a fantasized nature and rurality that pervades the names of suburbs and peripheral residential complexes. The spectacle displayed in the center of metropolises is the imaginary of a progressive and commodified world culture, inspired by world cities: "factories" and other "labs", "valleys", "Central Parks" or neighborhoods mimicking Soho with names made up of the initials of cardinal points. In this sense, a theoretical approach based on the notion of the spectacle makes it possible to go beyond or articulate the concepts of hegemony and biopolitics – in the sense of the conduct of conducts of the acting populations – as well as to connect political and market projects. More, this approach gives a political dimension to the questions of reception, appropriation and even subversion and performance. The infatuation with the figure of Mandela throughout the Black Atlantic, in the center and the extreme periphery of Northern and Southern metropolises, displays how incredibly diversified the toponymic projection of the contemporary globalized society of the spectacle is, and how this projection is able to coopt forms of self-promotion and forms of contestation by using a toned-down reference pointing to a desire for happy, even paternalistic harmony.

On the other hand, situationism allows for inventing situations for which alternative toponymy, as diverse as it can be, offers new horizons for struggle and play, which might include and empower.

Thus, Greil Marcus (2009), explaining the situationist project – that of urban drifting, psychogeography and subversion – implicitly reveals the possibility of a radical reinvention of the environment through words, in other terms, of the toponymic creation of situations:

> In this new world, the disconnected, seemingly meaningless words and pictures of Mémoires would make sense. They would make sense, first, as noise, a cacophony ripping up the syntax of social life – the syntax, as Debord put it in *The Society of the Spectacle*, of "the existing order's uninterrupted discourse about itself". As the noise grew, those words and pictures would begin to link up – as graffiti on countless walls, shouts coming out of thousands of mouths, even as familiar streets and buildings one suddenly saw as if never before – and then, with the old syntax broken, these things would make a second kind of sense. They would be experienced not as things at all, but as possibilities: elements of what Debord called "constructed situations" [...].
>
> Each situation would be an "ambient milieu" for a "game of events"; each would change its setting, and allow itself to be changed by it. The city would no longer be experienced as a scrim of commodities and power [...].
>
> When this free field was finally opened by the noise of the exploding syntax, when the fall of the dictionary left all words lying in the streets, when men and women rushed to pick them up and make pictures out of them, such daydreams would find themselves empowered, turning into catalysts for new passions, new acts, new events: situations, "made to be lived by their creators", a whole new way of being in the world. These situations would make a third kind of sense: they would seem sui generis, unencumbered by the baggage of any past, opening always into other situations, and into the new kind of history it would be theirs to make. And this would be a history not of great men, or of the monuments they had left behind, but a history of moments: the sort of moments everyone once passed

through without consciousness and the sort of moments everyone once passed through without consciousness and that, now, everyone would consciously create.

## 1.5. References

Alderman, D. (2008). Place, naming, and the interpretation of cultural landscapes. In *The Ashgate Research Companion to Heritage and Identity*, Graham, B. and Howard, P. (eds). Ashgate Press, Aldershot.

Almeida-Topor, H. (1996). Le modèle toponymique colonial dans les capitales de l'Ouest africain francophone. In *La Ville européenne outre mers : un modèle conquérant ?* Coquery-Vidrovitch, C. and Goerg, O. (eds). L'Harmattan, Paris.

Austin, J.L. (1963). *How to Do Things with Words*. Harvard University Press, Cambridge, MA.

Azaryahu, M. (1996). The power of commemorative street names. *Environment & Planning D: Society & Space*, 14(3), 311–330.

Azaryahu, M. (2011). The critical turn and beyond: The case of commemorative street naming. *ACME: An International Journal for Critical Geographies*, 10(1), 28–33.

Badariotti, D. (2002). Les noms de rue en géographie. Plaidoyer pour une recherche sur les odonymes. *Annales de Géographie*, 285–302.

Badariotti, D., Pfaender, F., Pinol, J.-L. (2021). Trente millions d'adresses : boulevards et impasses d'une recherche sur la complexité lexicale et spatiale des odonymes. *Cybergeo: European Journal of Geography*. DOI: https://doi.org/10.4000/cybergeo.36989.

Bancilhon, M., Constantinides, M., Bogucka, E.P., Aiello, L.M., Quercia, D. (2021). Streetonomics: Quantifying culture using street names. *PLoS One*, 17. DOI: https://doi.org/10.1371/journal.pone.0252869.

Beaudoin, S. and Martin, G. (2019). *Femmes et Toponymie. De l'occultation à la parité*. Éditions du Fleurdelysé, Sherbrooke.

Beck, L. (2021). Euro-settler place naming practices for north America through a gendered and racialized lens. *Terrae Incognitae*, 53(1), 5–25.

Berg, L.D. (2011). Banal naming, neoliberalism, and landscapes of dispossession. *ACME: An International Journal for Critical Geographies*, 10(1), 13–22.

Berg, L.D. and Kearns, R.A. (1996). Naming as norming: "Race", gender, and the identity politics of naming places in Aotearoa/New Zealand. *Environment and Planning D: Society and Space*, 14(1), 99–122.

Berg, L.D. and Vuolteenaho, J. (eds) (2009). *Critical Toponymies: The Contested Politics of Place Naming*. Ashgate Press, Aldershot.

Bertrand, M. (2001). Dynamiques urbaines, composition toponymique : le cas de Bamako (Mali). In *Nommer les nouveaux territoires urbains*, Rivière d'Arc, H. (ed.). Éditions de la Maison des sciences de l'homme – Editions de l'Unesco – Open Editions, Paris.

Bigon, L. (2008). Names, norms and forms: French and indigenous toponyms in early colonial Dakar, Senegal. *Planning Perspectives*, 23(4), 479–501.

Bigon, L. (ed.) (2016). *Place Names in Africa. Colonial Urban Legacies, Entangled Histories*. Springer, Cham.

Bigon, L. (2020). Towards creating a global urban toponymy – A comment. *Urban Science*, 4(4), 75. DOI: https://doi.org/10.3390/urbansci4040075.

Bigon, L. and Zuvalinyenga, D. (2021). Urban Pulse – Gendered urban toponyms in the global South: A time for de-colonization? *Urban Geography*, 42(2), 226–239.

Billig, M. (1995). *Banal Nationalism*. Sage, London.

Bing, J.-B. (2020). La variation toponymique dans l'oraliture comme pratique infrapolitique : études de cas à Uchon et Paris. *Echogéo*, 53. DOI: https://doi.org/10.4000/echogeo.19897.

Blackwood, R., Lanza, E., Woldemariam, H. (eds) (2016). *Negotiating and Contesting Identities in Linguistic Landscapes*. Bloomsbury Publishing, London.

Bourillon, F. (2012). *Changer les noms de rues de Paris. La Commission Merruau – 1862*. Presses universitaires de Rennes, Rennes/Comité d'histoire de la ville de Paris, Paris.

Bouvier, J.-C. (2007). Le langage de la toponymie urbaine. Approche méthodologique. *Onoma*, 42, 23–38.

Bouvier, J.-C. and Guillon, J.-M. (eds) (2001). *La toponymie urbaine. Significations et enjeux*. L'Harmattan, Paris.

Boyer, F. (2014). "Faire fada" à Niamey (Niger) : un espace de transgression silencieuse ? *Carnets de géographes*, 7. DOI: https://doi.org/10.4000/cdg.421.

Boyer, F. (2017). Les graffitis : une inscription spatiale et temporelle de la jeunesse dans l'espace public. *Urbanités*, 9 [Online]. Available at: https://www.revue-urbanites.fr/9-quand-les-fada-se-peignent-sur-les-murs-les-jeunes-hommes-dans-lespace-public-a-niamey-niger/.

Brasher, J.P., Alderman, D.H., Inwood, J.F.J. (2017). Applying critical race and memory studies to university place naming controversies: Toward a responsible landscape policy. *Papers in Applied Geography*, 1–16.

Caïazo, L. and Nick, I.M. (eds) (2020). *Shifting Toponymies: (Re)naming Places, (Re)shaping Identities*. Cambridge Scholars Publishing, Newcastle upon Tyne.

Cardoso, R.V. and Meijers, E.J. (2017). The metropolitan name game: The pathways to place naming shaping metropolitan regions. *Environment and Planning A*, 49(3), 703–721.

Carrie, K., Maggie, M., Michael, H. (2021). Rename and resist settler colonialism: Land acknowledgments and Twitter's toponymic politics. *First Monday*, 26(2). DOI: https://doi.org/10.5210/fm.v26i2.11454.

Carter, P. (1987). *The Road to Botany Bay*. University of Minnesota Press, Mineapolis, MN.

Casagranda, M. (2020). Re-naming and re-mapping Nunavut: An analysis of two digital maps. In *Shifting Toponymies: (Re)naming Places, (Re)shaping Identities*, Caiazzo, L. and Nick, I.M. (eds). Cambridge Scholars Publishing, Newcastle upon Tyne.

de Certeau, M. (1980). *L'invention du quotidien*. Gallimard, Paris.

Chamboredon, J.-C. (1988). Carte, désignations territoriales, sens commun géographique : les "noms de pays" selon Lucien Gallois. *Études rurales*, 109, 5–54.

Charbonneaux, J. (2020). Du toponyme à l'incarnation métonymique : la renomination parisienne de la place et de la station de métro "Europe" en "Europe-Simone Veil". *De Europa*, 3(2), 153–163.

Chaté, A. (2005). *Villa Mon Rêve, Ker Lulu : analyse sociologique des noms de maison*. L'Harmattan, Paris.

Chatelier, J. (2007). La révision toponymique (et cartographique) en Nouvelle-Calédonie (1983–1993). *Journal de la Société des Océanistes*, 125(2), 295–310.

Choplin, A. and Lozivit, M. (2019). Mettre un quartier sur la carte : cartographie participative et innovation numérique à Cotonou (Bénin). *Cybergeo: European Journal of Geography*. DOI: https://doi.org/10.4000/cybergeo.32152.

Cohen, S.B. and Kliot, N. (1981). Israel's place-names as reflection of continuity and change in nation-building. *Names: A Journal of Onomastics*, 29(3), 227–248.

Cohen, S.B. and Kliot, N. (1992). Place-names in Israel's ideological struggle over the administered territories. *Annals of the Association of American Geographers*, 82(4), 653–680.

Collignon, B. (1996). *Les Inuit : ce qu'ils savent du territoire*. L'Harmattan, Paris.

Cornips, L. and de Rooij, V.A. (eds) (2018). *The Sociolinguistics of Place and Belonging. Perspectives from the Margins*. John Benjamins Publishing Company, Amsterdam.

Cumbe, C. (2016). Formal and informal toponymic inscriptions in maputo: Towards socio-linguistics and anthropology of street naming. In *Place Names in Africa*, Bigon, L. (ed.). Springer, Cham.

Debarbieux, B. and Gumuchian, H. (1987). Représentations spatiales et dénominations des territoires : l'inscription toponymique des aménagements touristiques récents dans les Alpes du Nord (Savoie). *Revue de géographie alpine*, 2, 171–182.

Debord, G. (1967). *La Société du spectacle*. Buchet-Chastel, Paris.

Dorrier-Apprill, E. and Van den Avenne, C. (2002). Usages toponymiques et pratiques de l'espace urbain à Mopti (Mali) – La toponymie entre linguistique et géographie. *Marges linguistiques*, 3, 152–158.

Duncan, J. (1990). *The City as Text: The Politics of Landscape Inscription in the Kandyan Kingdom*. Cambridge University Press, Cambridge.

Düzgün, D. (2020). The death of the author in the streetscape of Ankara: A Barthesian intervention in critical toponymy. Thesis, Middle East Technical University, Ankara.

Eades, G.L. (2017). *The Geography of Names: Indigenous to Post-Foundational.* Routledge, Milton Park.

Fabiszak, M., Buchstaller, I., Brzezińska, A.W., Alvanides, S., Griese, F., Schneider, C. (2021). Ideology in the linguistic landscape: Towards a quantitative approach. *Discourse & Society*, 32(4), 405–425.

Gallois, L. (1908). *Régions naturelles et noms de pays.* Armand Colin, Paris.

Gardy, P. (1987). Réflexions à propos d'un "toponyme" : Larzac, le nom impropre ? *Cahiers de praxématique*, 8, 67–76.

Gensane, B. (2010). Penny Lane et Strawberry Field(s) : réel, identité et fantasmagorie. *Les Cahiers du MIMMOC*, 6. DOI: https://doi.org/10.4000/mimmoc.482.

Giraut, F. (2020). Dénominations plurielles : quand les noms de lieux se concurrencent ! *Echogéo*, 53. DOI: https://doi.org/10.4000/echogeo.20167.

Giraut, F. and Houssay-Holzschuch, M. (2008a). Au nom des territoires ! Enjeux géographiques de la toponymie. *L'Espace géographique*, 37(2), 97–105.

Giraut, F. and Houssay-Holzschuch, M. (2008b). Néotoponymie : formes et enjeux de la dénomination des territoires émergents. *L'Espace politique*, 2, 5–12.

Giraut, F.S., Guyot, S., Houssay-Holzschuch, M. (2008). Enjeux de mots : les changements toponymiques sud-africains. *L'Espace géographique*, 37(2), 131–150.

Giraut, F. and Houssay-Holzschuch, M. (2016). Place naming as dispositif: Toward a theoretical framework. *Geopolitics*, 21(1), 1–21.

Gnatiuk, O. and Glybovets, V. (2020). Do street status and centrality matter for post-socialist memory policy? The experience of Ukrainian cities. *Geographia Polonica*, 93(2), 139–161.

Gorter, D., Marten, H.F., Van Mensel, L. (2011). *Minority Languages in the Linguistic Landscape.* Springer, London.

Gottmann, J. (1952). *La politique des états et leur géographie.* Armand Colin, Paris.

Guichonnet, P. (1951). La toponymie savoyarde et les nouvelles cartes de l'Institut Géographique National. *Revue de géographie alpine*, 39(1), 201–211.

Guillorel, H. (ed.) (2008). *Toponymie et politique : les marqueurs linguistiques du territoire.* Bruylant, Brussels.

Guillorel, H. (2012). Onomastique, marqueurs identitaires et plurilinguisme. Les enjeux politiques de la toponymie et de l'anthroponymie. *Droit et culture*, 64, 11–50.

Hagen, J. (2011). Theorizing scale in critical place-name studies. *ACME: An International Journal for Critical Geographies*, 10(1), 23–27.

Hirschman, A.O. (1991). *The Rhetoric of Reaction*. Harvard University Press, Cambridge, MA.

Hopkins, J. (1998). Signs of the post-rural: Marketing myths of a symbolic countryside. *Geografiska Annaler: Series B, Human Geography*, 80(2), 65–81.

Houssay-Holzschuch, M. (2008). Nomen est omen. Lectures des changements toponymiques. *L'Espace géographique*, 37(2), 153–159.

Irvine, P.M., Memela, S., Dlongolo, Z.N., Kepe, T. (2021). Navigating community and place through colloquial street names in Fingo village, Makhanda (Grahamstown). *Urban Forum*, 32, 333–348.

Jenkins, E. (1990). From leliefontein to megawatt park: Some trends in the naming of South African places. *English Academy Review*, 7(1), 60–67.

Jenkins, E. (2007). *Falling into Place: The Story of Modern South African Place Names*. David Philip, Cape Town.

Jenson, J. (1995). Mapping, naming and remembering: Globalization at the end of the twentieth century. *Review of International Political Economy*, 2(1), 96–116.

Kadmon, N. (2000). *Toponymy: The Lore, Laws, and Language of Geographical Names*. Vantage Press, New York.

Karlander, D. (2017). Roads to regimentation: Place, authenticity and the metapragmatics of naming. *Language & Communication*, 53, 11–24.

Kasi-Tani, T. (2014) Le skateur comme designer. *Environnement Urbain /Urban Environment*, 8 [Online]. Available at: http://journals.openedition.org/eue/219.

Kavaratzis, M. (2004). From city marketing to city branding: Towards a theoretical framework for developing city brands. *Place Branding*, 1(1), 58–73.

Kleivan, I. (1977). Place names in greenland: Cultural imperialism and cultural identity. *Transactions of the Finnish Anthropological Society*, 2, 197–215[7].

Kripke, S.A. (1972). Naming and necessity. *Semantics of Natural Language*. Springer, Dordrecht.

Kristol, A.M. (2002). Motivation et remotivation des noms de lieux : réflexions sur la nature linguistique du nom propre. *Rives nord-méditerranéennes*, 11, 105–120.

Lajarge, R. and Moses, C. (2008). Néotoponymie, marqueur et référent dans la recomposition de territoires urbains en difficulté. *L'Espace politique*, 2, 97–111.

Laugier, S. (2004). Acte de langage ou pragmatique ? *Revue de métaphysique et de morale*, 42(2), 279–303.

Leimdorfer, F., Couret, D., Kouadio N'Guessan, J., Soumahoro, C., Terrier, C. (2002). Nommer les quartiers d'Abidjan. In *Les divisions de la ville*, Topalov, C. (ed.). Éditions de la Maison des sciences de l'Homme, Paris.

7. Article pointed out by Marine Duc.

Lewis, P.G. (1982). The politics of Iranian place-names. *Geographical Review*, 72(1), 99–102.

Light, D. and Young, C. (2014). Toponymy as commodity: Exploring the economic dimensions of urban place names. *International Journal of Urban and Regional Research*, 39(3), 435–450.

Light, D., Nicolae, I., Suditu, B. (2002). Toponymy and the communist city: Street names in Bucharest, 1948–1965. *GeoJournal*, 56(2), 135–144.

Marcus, G. (2009). *Lipstick Traces. A Secret History of the Twentieth Century, Twentieth Anniversary Edition*. Harvard University Press, Cambridge, MA.

Medway, D. and Barnaby, G. (2014). What's in a name? Place branding and toponymic commodification. *Environment and Planning A*, 46(1), 153–167.

Mill, J.S. (1896). *Système de logique déductive et inductive*, 4th edition, translated by L. Peisse. Alcan, Paris.

Milo, D. (1984). Les noms de rues. In *Les lieux de Mémoire*, Nora, P. (ed.). Gallimard, Paris.

Moeschler, J. (2019). Pragmatique des noms propres (et des noms de lieux) [Online]. Available at: https://neotopo.hypotheses.org/1960.

Mokwena, L. (2021). Blurring the boundaries between "urban" and "rural": Oral route directions and the invention of street names in Ulco and Delportshoop. *South African Geographical Journal*, 103(3), 303–318.

Monmonier, M.S. (2006). *From Squaw Tit to Whorehouse Meadow: How Maps Name, Claim, and Inflame*. University of Chicago Press, Chicago, IL.

Montès, C. (2008). La toponymie comme révélateur de la construction identitaire d'un empire : (re)nommer les capitales étatsuniennes. *L'Espace géographique*, 37(2), 106–116.

Müller-Wille, L. (1989). Place names, territoriality and sovereignty: Inuit perception of space in Nunavik (Canadian Eastern Arctic). *Schweizerische Amerikanisten-Gesellschaft Bulletin*, 53(54), 1990.

Nadiras, S. (ed.) (2018). *Noms de lieux, noms de personnes : la question des sources*. Publications des Archives nationales, Pierrefitte-sur-Seine.

Nash, C. (1993). Remapping and renaming: New cartographies of identity, gender and landscape in Ireland. *Feminist Review*, 44, 39–57.

Nash, J. (2015). The how of toponymy: A comment on Tent's "Approaches to Research in Toponymy". *Names*, 63(4), 233–236.

Ndemo, B. (2020). Slum digitisation, its opponents and allies in developing smart cities: The case of Kibera, Nairobi. *Open Cities, Open Data*, 129–148.

Ndock Ndock, G. (2020) Urbanisme de rattrapage, marquage territorial populaire et conflits d'odonymies dans les quartiers de Yaoundé (Capitale du Cameroun). *Echogéo*, 53. DOI: https://doi.org/10.4000/echogeo.20168.

Neotopo (2014). Penny Lane comme métatoponyme [Online]. Available at: https://neotopo.hypotheses.org/30.

Neotopo (2020). Lascaux : du lieu-dit à la marque globale [Online]. Available at: https://neotopo.hypotheses.org/2385.

Njoh, A.J. (2010). Toponymic inscription, physical addressing and the challenge of urban management in an era of globalization in Cameroon. *Habitat International*, 34(4), 427–435.

Njoh, A.J. (2017). Toponymic inscription as an instrument of power in Africa: The case of colonial and post-colonial Dakar and Nairobi. *Journal of Asian and African Studies*, 52(8), 1174–1192.

Norris, D.A. (1999). Unreal estate: Words, names and allusions in suburban home advertising. *Names: A Journal of Onomastics*, 47(4), 365–380.

Novas Ferradás, M.A. (2018). Commemorative urban practices and gender equality: The case of Santiago de Compostela's urban anthroponymy. *Hábitat y Sociedad*, 11, 109–129.

Oto-Peralias, D. (2018). What do street names tell us? The "city-text" as socio-cultural data. *Journal of Economic Geography*, 18(1), 187–211.

Palmer, M. and Corson, C. (2020). Decolonizing world heritage maps using indigenous toponyms, stories, and interpretive attributes. *Cartographica: The International Journal for Geographic Information and Geovisualization*, 55(3), 183–192.

Palonen, E. (2008). The city-text in post-communist Budapest: Street names, memorials, and the politics of commemoration. *GeoJournal*, 73(3), 219–230.

Paunonen, H., Vuolteennaho, J., Ainiala, T. (2009). Industrial urbanization, working-class lads and slang toponyms in early twentieth-century Helsinki. *Urban History*, 36(3), 449–472.

Paveau, M.-A. (2008). Le toponyme, désignateur souple et organisateur mémoriel. L'exemple du nom de bataille. *Mots*, 86(1), 23–35.

Perkins, H.C. (1989). The country in the town: The role of real estate developers in the construction of the meaning of place. *Journal of Rural Studies*, 5(1), 61–74.

Pirie, G.H. (1984). Letters, words, worlds the naming of Soweto. *African Studies*, 43(1), 43–51.

Pourtier, R. (1983). Nommer l'espace. Émergence de l'État territorial en Afrique noire. *L'Espace géographique*, 12(4), 293–304.

Puzey, G. and Kostanski, L. (eds) (2016). *Names and Naming: People, Places, Perceptions and Power*. Multilingual Matters, Bristol.

Puzey, G. and Vuolteenaho, J. (2016). Developing a Gramscian approach to toponymy. Names and their environment. *Proceedings of the 25th International Congress of Onomastic Sciences*, Glasgow.

Raffestin, C. (1986). Écogenèse territoriale et territorialité. In *Espaces, jeux et enjeux*, Auriac, F. and Brunet. R. (eds). Fayard, Paris.

Randall, R.R. (2001). *Place Names: How They Define the World – and More*. Scarecrow Press, Lanham.

Recanati, F. (1983). La sémantique des noms propres : remarques sur la notion de “désignateur rigide”. *Langue française*, 57, 106–118.

Reynaud, A. (1990). La Champagne : le mot et la chose. Un nom à la recherche d'un espace. *Travaux de l'Institut de Géographie de Reims*, 5–13.

Ritchot, G. (1989). Géographie structurale et toponymie contradictoire. *Cahiers de Géographie du Québec*, 33(88), 67–71.

Rodman, W.L. and Rodman, M. (1985). Rethinking kastom: On the politics of place naming in Vanuatu. *Oceania*, 55(4), 242–251.

Rose-Redwood, R. (2011). With numbers in place: Security, territory, and the production of calculable space. *Annals of the Association of American Geographers*, 102(2), 295–319.

Rose-Redwood, R. (2016). “Reclaim, rename, reoccupy”: Decolonizing place and the reclaiming of PKOLS. *ACME: An International Journal for Critical Geographies*, 15(1), 187–206.

Rose-Redwood, R. and Alderman, D. (eds) (2011). New directions in political toponymy, *ACME: An International Journal for Critical Geographies*, 10(1), 1–6 [Online]. Available at: https://acme-journal.org/index.php/acme/article/view/879.

Rose-Redwood, R., Alderman, D., Azaryahu, M. (2010). Geographies of toponymic inscription: New directions in critical place-name studies. *Progress in Human Geography*, 34(4), 453–470.

Rose-Redwood, R., Alderman, D., Azaryahu, M. (eds) (2018). *The Political Life of Urban Streetscapes*. Routledge, London.

Rusu, M.S. (2020). Street names through sociological lenses. Part I: Functionalism and conflict theory. *Social Change Review*, 18(1), 1–24.

Rusu, M.S. (2021). Street names through sociological lenses. Part II: Constructionism and utilitarianism. *Social Change Review*, 19(1), 1–29.

Scott, J.C. (1998). *Seeing Like a State: How Certain Schemes to Improve the Human Condition Have Failed*. Yale University Press, New Haven, CT.

Searle, J.R. (1963). Proper names. In *Philosophy and Ordinary Language*, Caton, C.E. (ed.), University of Illinois Press, Urbana, IL.

Searle, J.R. (1969). *Speech Acts: An Essay in the Philosophy of Language*. Cambridge University Press, Cambridge.

Smith, B. (2018). Engaging geography at every street corner: Using place-names as critical heuristic in social studies. *The Social Studies*, 109(2), 112–124.

Steffens, S. (2007). La toponymie populaire urbaine hier et aujourd'hui. Le cas de Molenbeek-Saint-Jean. *Brussels Studies*. DOI: https://doi.org/10.4000/brussels.441.

Stewart, G.R. (1945). *Names on the Land: A Historical Account of Place-Naming in the United States*. Random House, New York.

Stickler, P.J. (1990). Invisible towns: A case study in the cartography of South Africa. *GeoJournal*, 22(3), 329–333.

Tent, J. (2015). Approaches to research in toponymy. *Names*, 63(2), 65–74.

Tent, J. and Blair, D. (2011). Motivations for naming: The development of a toponymic typology for Australian placenames. *Names*, 59(2), 67–89.

Traore, B. (2007). Toponymie et histoire dans l'Ouest du Burkina Faso. *Journal des africanistes*, 77(1), 75–111.

Tuan, Y.-F. (1991). Language and the making of place: A narrative-descriptive approach. *Annals of the Association of American Geographers*, 81(4), 684–696.

Vaneigem R. (1967). *Traité de savoir-vivre à l'usage des jeunes générations*. Gallimard, Paris.

de Vergottini, G. and Piergigli, V. (eds) (2011). *Topographical Names and Protection of Linguistic Minorities*. Peter Lang, Bern.

Veschambre, V. (2011). La mise en mots des espaces dans un grand ensemble en rénovation. Décalages significatifs entre toponymie conçue et toponymie vécue (Malakoff, Nantes, France). *Cahiers internationaux de sociolinguistique*, 1(1), 99–110.

Vivant, E. (2020). Nommer les lieux de la crise des opioïdes à Boston : un enjeu politique. *Echogéo*, 53. DOI: https://doi.org/10.4000/echogeo.20001.

Vuolteenaho, J. and Kolamo, S. (2012). Textually produced landscape spectacles? A Debordian reading of Finnish namescapes and English soccerscapes. *COLLEGIUM: Studies across Disciplines in the Humanities and Social Sciences*, 13, 132–158.

Vuolteenaho, J. and Puzey, G. (2018). Armed with an encyclopedia and an axe: The socialist and post-socialist street toponymy of East Berlin revisited through Gramsci. In *The Political Life of Urban Streetscapes: Naming, Politics and Place*, Rose-Redwood, R., Alderman, D., Azaryahu, M. (eds). Routledge, Abingdon.

Vuolteenaho, J., Lappalainen, H., Ainiala, T. (2019). A slightly slummier area? Negotiations of place-bound identities through social spatializations and unofficial toponyms. *Journal of Sociolinguistics*, 23(2), 140–162.

Wanjiru, M.W. and Matsubara, K. (2017). Slum toponymy in Nairobi, Kenya. A case study analysis of Kibera, Mathare and Mukuru. *Urban and Regional Planning Review*, 4, 21–44.

Włoskowicz, W. (2019). Dissemination and correctness of geographical names. Geographers and cartographers as toponym users and promoters. *Annals of the Austrian Geographical Society*, 161, 291–324.

Wonders, W.C. (1987). Native claims and place names in Canada's western arctic. *The Canadian Journal of Native Studies*, 7(1), 111–120.

Wood, D. (2002). Selling the suburbs: Nature, landscape, adverts, community. *Transformations*, 5 [Online]. Available at: http://www.transformationsjournal.org/wp-content/uploads/2017/01/denniswood.pdf.

Wright, L. (2020). *Sunnyside: A Sociolinguistic History of British House Names*. Oxford University Press, Oxford.

Xu, W. (2021). Can one artist name an unnamed road after himself? *Cultural Geographies*, 1–7 [Online]. Available at: https://doi.org/10.1177/14744740211020509.

Yeoh, B.S.A. (1996). Street-naming and nation-building: Toponymic inscriptions of nationhood in Singapore. *Area*, 28(3), 298–307.

Zelinsky, W. (1967). Classical town names in the United States: The historical geography of an American idea. *Geographical Review*, 57(4), 463–495.

Zelinsky, W. (1983). Nationalism in the American place-name cover. *Names*, 31(1), 1–28.

Zuvalinyenga, D. and Bigon, L. (2021). Gender-biased street naming in urban sub-Saharan Africa: Influential factors, features and future recommendations. *Journal of Asian and African Studies*, 56(3), 589–609.

# 2

# Commemorative Place Naming: To Name Places, to Claim the Past, to Repair Futures

**Derek H. ALDERMAN**
*Department of Geography & Sustainability, University of Tennessee, Knoxville, USA*

## 2.1. A renaming moment in Paris

Activists in the French feminist organization, *Osez le féminisme*, captured news headlines in 2015 when they unofficially renamed 60 streets in Paris. They covered existing road signs with the names of women from history, including notable singers, lawyers, scientists and physicians. The protest was meant to draw attention to the scarcity of Parisian roadways honoring women: little over 7% of streets named after women (male names represent 51% of all streets) and most of those commemorated females are saints or the wives or daughters of well-known men. Some observers characterized the guerilla renaming campaign as merely a publicity "stunt", but *Osez le féminisme* saw it as a strategy for provoking Paris' mayor to discuss a "concrete action plan" for bringing gender parity in place naming (Gee 2015).

This rewriting of Paris street names is not an isolated event. Sexism in commemorative naming is found across many world cities (Poon 2015; Forrest 2018; Mamvura et al. 2018), and campaigns in other countries are pushing for greater landscape representation of female historical figures. Activists are also challenging long-standing place names (or toponyms) that valorize historical figures associated with racism and imperialism and the lack of racial and ethnic diversity in commemoration. Contentious debates are being waged across regions over who and what should be memorialized – not only on street signs, but also in the names of schools, parks, university buildings and other public places (Brasher et al. 2017).

The world is the midst of a renaming moment. Place name change is increasingly pursued with the hope of inscribing new histories, identities and more inclusive social values into landscapes. Toponyms are being used by activists as reparative tools for recognizing and doing justice to erased indigenous ties to the land and the neglected struggles of people of color, women and queer communities. These changes grow out the "adamant assertions of citizen rights and persistent demands for representation and respect" by public groups (Doss 2010, p. 2). While the renaming moment is ostensibly a struggle over the past, it is also a struggle over whose identities and lives matter now and into the future – especially for those traditionally marginalized.

Memorial toponyms represent highly charged civic issues within communities, but many elected officials, journalists, citizen groups and even some scholars lack a full understanding of these struggles and what they mean. Although often taken for granted, commemorative place naming is important to people's storytelling, lived material experiences and political–emotional well-being as they inhabit, claim and create places. To understand the protests in Paris and elsewhere, this chapter identifies the commemorative work underlying and performed by place naming. We review established and emerging approaches within the field to explain the narrative, affective and material capacities of memorial toponyms along with their reparative possibilities and limits.

## 2.2. Place naming as commemorative work

Commemorative place naming refers to a self-conscious invocation of the past, often of the memory of historical people and events. But it is worth noting that all toponyms, even those not explicitly devoted to memorialization, are invested with a sense of history and serve as storehouses of personal and social memories and experiences that guide people's actions and identification (Rose-Redwood 2008). To comprehend fully the relationship between commemoration and place naming, we must recognize the socially constructed nature of referencing and remembering history. The vision of history depicted through the memorial landscape – which includes toponyms along with museums, monuments, historical performances and preserved sites – is not synonymous with all that has happened in the past (Nora 1989). What is defined as memorable is socially mediated and the result of commemorative work and decisions of individuals and groups – from artists, tour guides and community activists to government officials, museum curators, business leaders – who as "memorial entrepreneurs" influence how the public views, senses and debates the past (Naef 2019). Commemoration is obviously about the past but is also situated within the present as social actors and groups control, negotiate and contest memory to serve contemporary cultural, economic and political needs (Dwyer and Alderman 2008). Named places are the product of the work of memorialization and in turn produce or perform work within society.

Place naming gives tangibility and familiarity to ideas and beliefs about the past, normalizing memory and allowing it to appear as historical and social truth. While many people might treat place names as impartial recorders or markers of history, commemoration is an inherently selective practice that can hide as much as it reveals (Bouvier 2007). Remembering is accompanied, simultaneously, by a process of forgetting – an excluding of other historical narratives and identities from public consideration. Because toponyms often reflect the values and worldviews of dominant, elite social classes, they tend to ignore the past experiences and struggles of marginalized or subaltern groups. Yet, drawing a hard binary between elite and marginalized identities does not fully capture all commemorative tensions and interests vying for control of the past and place (Rose-Redwood 2008).

Behind every study of toponymic commemoration should be a series of major questions. What political actors, decisions and social relations enable certain memorialized names to be displayed publicly? Who is given the power and authority (or not) to do this work of remembering and naming? Whose identifications with the past are enacted (or denied) through place naming? How do these toponyms work to benefit (or disadvantage) the belonging and social standing of certain public groups over others?

The tension between remembering and forgetting found in named places never goes unchallenged, in part because heritage is inherently dissonant or open to disagreement (Tunbridge and Ashworth 1996). Plus, memory is more than what is intentionally commemorated; it also emerges involuntarily or unexpectedly when the past haunts the present (Atencio 2018). Even when certain histories are not written into memorial landscapes, this forgetting of the past can exert an "'absent presence" that affects the feel of places and thus calls attention to what and who is missing (Muzaini 2015). For *Osez le féminisme*, for example, social memories of women were conspicuously absent on named roadways. Activists used the counter-naming of streets to make visible who had long been invisible and to remind us of the "political choices" behind Parisian toponyms and more generally behind street names in all large European cities (Gee 2015).

Places of commemoration are in a constant state of becoming despite their apparent permanence (Jones and Garde-Hansen 2012). Toponyms are open to control, challenge and sometimes significant change as history is re-evaluated and re-purposed in response to market trends, ideological shifts and political struggles. Commemoratively named places have a projective power; they help create an historical frame of reference and sense of place for people that can structure, legitimize and reproduce social relations. Memorial toponyms – along with all place naming – function as a "political technology" used to order, govern or even resist the material and symbolic construction of places and who comes to count or matter in those places (Rose-Redwood et al. 2018). Controlling the representation and performance of the past as a socio-political resource is important to acquiring the symbolic capital or social distinction frequently at the heart of many power-laden identity struggles (Duminy 2014). How we imagine and portray ourselves in the contemporary and the future is intimately linked to how we name, remember and represent ourselves in relation to the past, providing justification for why we should be recognized and respected publicly.

## 2.3. Narrative capacities

Memorial toponyms are semiotic texts embedded in larger systems of representation, discourse and storytelling that help public groups make moral and ethical judgments about places, people and histories (Azaryahu 1996). Place naming is "rich in narrativity", even as it differs from other forms of commemorative communication (Ryan et al. 2016, p. 141). Place names consist of a limited number of words and cannot depict the past in the same detailed ways as museum exhibits or monument inscriptions – although contextual signage is sometimes attached to named features to expand their interpretive capacity. The toponym is like "the title of a story that stands for and encapsulates the life story of a person or an account of an event" (Ryan et al. 2016, p. 143). It serves as a powerful symbolic shorthand or signpost for wider stories about the past without being able to sufficiently unpack these histories.

Place naming is intertextual in nature, defined in part by how it works with or against master-narratives of national, regional and local history. Toponyms are not static vehicles for narratives but in a state of doing; they participate in *narrating* commemorative messages and converting memories into public historical knowledge. Named spaces are limited in fully narrating commemorated pasts, but they inscribe histories into the texts, performances and spaces of everyday life in ways not possible through museums, statuary and heritage tourism. Memorial toponyms, especially street names, incorporate remembered histories within people's sensory-laden movements and landmarks in such intimate ways that people are not fully cognizant of the memorial choices and impacts going on around them.

Because of the daily pervasiveness of place names, authorities often use them to narrate and legitimate ideas of national heritage and identity. Scholars find toponyms valuable for understanding how the (re)writing of collective memory is shaped by political transitions and revolutions, such as post-colonialism, post-socialism and post-apartheid (e.g. Giraut et al. 2008; Bigon 2009; Desbrosse 2014; Wanjiru and Matsubara 2017). It is common for government officials and other elites to remove toponyms and other memorials that symbolize earlier regimes, replacing them with names that sanctify a new set of heroes, military campaigns and ideological causes. This renaming of national history and tradition does more than alter texts, maps and signs. It is a consequential re-scripting of people's everyday lives, identities and associations and can provoke activism from above and below (Capdepón 2020).

Toponyms are not just the products of social power but also tools for claiming and maintaining certain power relationships (Giraut and Houssay-Holzschuch 2016). Control of language, especially in highly visible settings, is central to civic authority. Feminist activists in Paris – as the city's planners had long done (Bourillon 2016) – used commemorative place naming to fashion an alternative "political grammar", appropriating words and associated signs and memories to redefine the nation and who is included publicly as a citizen (Wilce 2012). A spokeswoman for *Osez le féminisme* explained the vision of nation-building behind their street name take over: "Little kids walking around Paris will subconsciously be taking in the history of France through things like street signs. They'll think that France was built by great men – but it's important they know about the important women too" (Gee 2015).

No matter how tightly we control the commemorative narratives communicated through toponyms, they can always be read, interpreted and acted upon in different ways by public groups, even in ways that conflict with the intent of officials in charge of naming (Azaryahu 2011). As Dwyer and Alderman (2008, p. 174) asserted: "No memorial speaks for itself; each one is dependent upon its audience to voice – or betray – its vision of the past into the future." Audiences form interpretive communities that vary based on their identities, personal memories, social experiences of privilege or marginality, and "narrativized worlds" – their pre-existing understandings of how the world functions and what it means to them (Carter et al. 2014). While place names have a capacity to spur new connections with the past, public groups also evaluate how memorialized names fit, positively or negatively, within a standard story plot long carried and deployed in their lives. It is perhaps more accurate to conceptualize toponym users as not just audiences but also co-authors of the meaning of memorialized names.

Ordinary residents can have strong attachments to the heritage of old, existing place name memorials, and thus resist or ignore government dictated rewriting of toponyms. Embracing ethnographic rather than just archival methods, researchers are paying more attention to popular responses to toponymic change. Practical concerns and ideological interests interlock in different ways for different citizens to shape their reception and use of new names and memories (Creţan and Matthews 2016; Vuolteenaho et al. 2019). Toponyms are embedded within people's daily habits, languages and corporeal enactments – including how they speak about named places (Kearns and Berg 2002; Light and Young 2014). Even a subconscious verbalizing of toponyms can act as a political utterance that mediates the process of remembering (or forgetting) the past. Inspired by more-than-representational approaches, scholars recognize that a sheer focus on the official spatial inscription process fails to recognize how memorial toponyms represent performative spaces (Rose-Redwood 2014). Analyzing the storytelling capacities of commemorative place naming involves more than a detached reading or listening. It also requires being sensitive to the embodied, multi-sensory ways that people experience and dwell within and beyond the narrated past of toponyms.

## 2.4. Affective capacities

A narrative analysis of place names is valuable but it does not fully capture the affective capacity of naming and the role of emotions in heritage (Tolia-Kelly et al. 2016). Memorial toponyms – as they become entangled with people's experiences, identities and social relations – can move public groups to feeling and action. They influence the public by participating in spectacular displays and performances of memory (e.g. parades, dedications, festivals, etc.), but their affective power is often non-spectacular or banal. This banality allows the memorialized name to appear ideologically innocent when in fact it is politically situated. By working unsuspectingly in the background, toponyms have the potential to create unexpected engagements with commemorated history and trigger a person's prior knowledge and memories – which could be comforting or painful. Sometimes, toponyms serve as cues that may inspire reflection and education about a past unfamiliar to people, a particularly important impact since many citizens are unaware of the history associated with names (Ryan et al. 2016).

Commemoratively named places affect public groups not only through direct history telling or memory-evoking but also by creating broader "commemorative atmospheres" (Sumartojo 2016). Place naming works with sensory experience, individual memory and the built environment to influence the feel, mood and ambience of the past. While the affective atmospheres of memorialized toponyms can reaffirm hegemonic narrations of the past, such as nationalistic heritage, they also can de-stabilize dominant ideas about memory, identity and place. In elevating the names and reputations of specific historical women on Paris streets, *Osez le féminisme* sought to create a disruptive commemorative atmosphere that would provoke public reaction that otherwise would not happen. In fact, a member of the feminist organization said she hoped the affected public would "realize and talk about it [street renaming], they will talk to their friends, maybe they'll start understanding a bit more" (Gee 2015).

An important but still undeveloped avenue of scholarship explores how the commemorative atmospheres of naming affect the socio-psychological well-being of communities, including the extent to which memorial toponyms mark and reproduce political and emotional violence (Alderman and Rose-Redwood 2020). While toponyms can evoke comfort, pride and belonging for one group, they can conjure feelings of alienation and intergenerational memories of trauma among other groups with histories of victimization and exclusion (Brasher et al. 2020). Place names are capable of perpetuating environmental micro-aggressions, those daily place-based indignities that communicate (whether intentionally or not) hostile, derogatory or negative slights and insults towards marginalized groups and thus reinforce institutional discrimination (Ferguson 2019). Important to protests in Paris, *Osez le féminisme* members viscerally sensed and resisted the micro-aggressions against women that had long been inscribed into the city's masculinist landscapes.

The term micro-aggressions can be misleading since the violent histories of place naming are not at all small. Toponymic practice was deeply rooted, in a systemic way, within the linguistic, physical and structural violence of settler colonialism and the creation of a commemorative tradition still widely seen and felt today. In that tradition, toponyms signify and legitimize the memories, identities and presumed belonging of White male settlers by publicly dispossessing indigenous people and communities of color of their own agency, histories and naming-language systems (Murphyao and Black 2015). The result of these naming patterns is a "memoricide" (Masalha 2015) that symbolically and materially annihilates and misrepresents subaltern attachments to place along with their historical subjectivities.

When toponyms perpetuate a public amnesia of the lived histories, experiences and feelings of oppressed groups, the larger society is not encouraged to think critically about what or who is missing from prevailing landscape expressions and how the excluded are affected. Toponyms and their attendant political-commemorative messages and atmospheres facilitate social exclusions in immediate and direct ways, but there is also a "slow" or attritional violence that is produced in shaping how we see ourselves and our histories, identities and political rights over generations – even to the point that some members of marginalized groups may not be aware of these effects (Ward 2015).

The southeastern USA is instructive when studying the violent aspects of memorial toponyms. There is a long history in the region of White supremacist social orders using representations of heritage to intimidate African Americans in their struggle for freedom and equality. The naming of places such as schools, streets and parks for racist historical figures such as slave-owners, Confederate leaders, members of the Ku Klux Klan and segregationist politicians was part of the same political-emotional order that sought to lynch and segregate Black communities and keep them from voting in elections and being employed in certain jobs (Brasher et al. 2020). The fixing of these racially charged names and memories onto the landscape has been instrumental rather than coincidental to White opposition to people of color making civil rights gains. For example, in direct protest of the 1954 Brown v. Board of Education Supreme Court decision that racially de-segregated public schools, officials in Montgomery, Alabama, named a high school for Robert E. Lee – the mythologized major-general of the pro-slavery, secessionist Confederate army during the American Civil War (Johnson 2019). Activists who have sought to remove these hostile toponyms are actively motivated by their own organic, day-to-day belief that these surviving racist symbols do affect them negatively, represent a form of discrimination and mark places "wounded" by the legacies of exclusion, discrimination and trauma (Till 2012).

## 2.5. Material capacities

Toponyms have a materiality that structures and shapes how people interact and identify with the memorialized past and the associated name. Place naming is part of the built environment as well as the realm of language. The very signage used to mark toponyms – while often perceived as mundane and merely informative – "constitute[s] part of a highly invested political strategy for producing a linguistic landscape" and maintaining or challenging symbolic power (Bigon and Dahamshe 2014, p. 606). Memorial toponyms work materially to make buildings and streets usable and recognizable; facilitate commerce, transportation and travel; participate in state formation and international politics; and demarcate zones of exclusivity and stigma in neighborhoods (Mitchelson et al. 2007; Adebanwi 2012; Spocter 2018). Place names participate in producing distinctive scalar constructions of memory that can reinforce long-standing socio-spatial boundaries or create new material connections between people, places and histories (Alderman and Inwood 2013).

Because the focus of traditional toponymic research has been on the name or historical reference itself, scholars have not fully analyzed the material spaces created around and through memorialization and naming.

Place names help form everyday infrastructures, understood here not as ideologically neutral physical surroundings but as consequential geographies that mediate broader distributions of power and access to opportunities, resources and rights. *Osez le féminisme* chose to re-identify streets in central Paris exactly because roads and their signage are a vital piece of infrastructure important to daily movements, activities and a sense of order. Coinciding with the 45th anniversary of Women's Liberation Movement, their toponymic protests challenged material inequalities that still constrain female claims on public space as well as public memory.

The central material position of commemorative place naming within the flows, practices and protests of people prompts us to consider its relational nature – how toponyms are embedded within broader and dynamic contexts, processes and relationships across time and space. Such a perspective moves away from the traditional approach of studying toponyms as single, isolated artifacts or linguistic patterns with a singular and stable meaning. Instead, commemoratively named spaces form and are formed by assemblages that toponymic practices share with a range of heterogeneous social actors and identities, politicized histories, cultural practices, government policies and lived spaces that work together and against each other to construct and transform the meaning of place and the past. The past assumes material expression and power specifically because of the relations that place naming forms with larger, fluid discourses, processes and materialities. For example, *Osez le féminisme*'s street name protest is inseparable from the organization's wider praxis in France that addresses the material expression of gender inequality – from the preservation of family planning to ending homophobia and violence against women. An assemblage perspective helps us realize the emergent and multiple ways in which places become named and memorable along with how discursive and material elements of toponyms assemble to spatialize history and historicize space (Wideman and Masuda 2018).

Important to the place naming assemblage is the relationship that commemoration has with location (Dwyer and Alderman 2008). Where public commemorative practices are emplaced within the material landscape is not incidental. The geographic configuration of toponyms creates a "spatial narrative" that affects how social actors and groups interact with and interpret the commemorated past (Azaryahu and Foote 2008). The material location of a place name affects the visibility, accessibility and symbolic prominence of the history it memorializes – all of which contributes to the degree to which the past resonates or appears relevant to the various social groups that constitute the public.

The resonance of a toponym is relationally defined and depends upon its proximity to other memorial locations, business districts, government centers, flows of locals and tourists, and broader geographies of privilege and marginality. This locational context contributes to the commemorative atmosphere that envelopes the memorial toponym, which can lend credibility to or evoke criticism of the past being commemorated. Some public struggles over whether to rename places are not just debates about who or what should be remembered but where exactly that commemorations should happen. Public groups readily recognize how the central or peripheral location of a toponym enhances or harms the legacy of the historical figure or event being remembered (Alderman and Inwood 2013).

Important to this focus on materiality is the political economy of memory and the commodification of toponymic practice (Karimi 2016). Influential donors and corporations increasingly purchase the right to affix their names to streets, stadia, university buildings, thus fundamentally reframing local place attachments and the identity of public spaces (Rose-Redwood et al. 2019). This place branding can lead to a politics of memory dispossession in which long-standing names and symbols of local heritage value are replaced along with possibly suppressing alternative commemorative messages of greater value to communities (Kearns and Lewis 2019). Memories, names and places take on an exchange value that overrides their use value for ordinary citizens, although the public can express resistance, acceptance or indifference to the commercialization (Medway et al. 2019).

Even when naming rights are not sold directly, toponyms can become part of a wider neo-liberal conversion and protection of the landscape for capital accumulation. Promoters often engineer the marketing image of housing developments, urban districts, medical centers, waterfronts, tourism destinations, commercial establishments and entire communities by deploying ambiguous, if not contrived historical symbols and names or wrapping older toponyms in a blanket of nostalgia (Madden 2018). While developers openly invent traditions through naming, they can also be reluctant to abandon long-time commemorative names – no matter how antiquated and offensive they may be – because of not wanting to disrupt the flow of capital, to inconvenience customers or to violate the prevailing rights of business and property owners affected by a change.

Manipulating and protecting memorial toponyms for capitalistic gain is often rationalized as non-political, although it clearly has ideological consequences and weakens the political will of government officials to address controversial or even the violent effects of established commemorative names. When such name-memory formations are eventually seen as a threat to capital accumulation, toponymic changes can be superficial, tokenistic and sanitizing to facilitate investment rather than truly addressing historical tensions and injustices (Brasher et al. 2020). Running through memorial toponyms is a dissonance between their way-finding and financial function and the political-commemorative work that activists hope to accomplish. French protesters exposed tension and temporarily challenged capitalist material power over Paris streets. They created a commemorative landscape where the educational and political message of toponyms took priority over doing "business as usual". *Osez le féminisme* appeared to relish the confusion their guerilla street renaming caused among tourists who reportedly "were standing on street corners consulting their maps" and "many [were] asking for directions" (Gee 2015).

## 2.6. Reparative possibilities and limits

No analysis of the Paris scene is complete without recognizing the extent to which toponyms serve as "memorial arenas", sites where social groups debate the meaning of history and compete for control over the commemorative process as part of larger struggles over identity and self-determination (Alderman 2002a). Subaltern communities and their supporters increasingly appropriate toponymic geographies as tools of resistance, but non-elite interpretations and support of counter memorial naming are not monolithic and social divisions within marginalized population can create inequalities in resistant toponymies (Forrest 2018; Brocket 2021). Some resistance takes place, as seen in France, through temporary toponymic protests. But historically oppressed groups often pressure their governments for official, permanent commemorative name changes as part of a broader exercising of the right to participate in public decision-making and the right to claim access to public spaces and symbols (Alderman and Inwood 2013). A fight for epistemic justice motivates much of this name reform as activists encourage the public to unlearn exclusionary ways of knowing the past and to learn new shared inclusive histories.

Memorial toponyms are arenas for "reputational politics", intense debates over the historical legacies and reputations of commemorated figures and the social connotations and appropriateness of memorializing them (Alderman 2002b). These debates lead to calls to remove long-standing place name memorials to historical figures known for perpetuating slavery, imperialism, genocide, antisemitism, eugenics, gender and sexual inequalities. Activist movements also want authorities to create new toponyms to recognize the historical contributions and civil rights struggles of women, people of color, indigenous communities, LGBTQ communities and members of other oppressed groups. This activism is about using public memory to expand the politics of who belongs or matters socially. In this respect, public memory is not just another word for social or collective memory, but a "process by which a group of people who were once dismissed and never thought of as part of a 'public' might become visible to themselves and to others" as "new publics" (Ebron 2014, p. 147).

Some marginalized communities and their allies believe the renaming of commemorative toponyms – because it enacts new political histories and new ideas of cultural citizenship – can serve as a form of "symbolic reparation" and assist in "restoring the dignity and public identity of victims" of oppression and rehabilitating national histories so that human rights violations are recognized (Swart 2008, p. 106). While reparation is often defined in strict monetary or litigation terms, a broader approach recognizes the role of commemoration in making amends for historical injustices and moving towards a more just future. The term "memory-work" (Till 2012) is of growing importance to capture the process of using new forms of public memory or counter-memorials, including place renames, to confront traditionally exclusionary commemorative structures and the broader continuing legacies of state-sanctioned discrimination, trauma and violence (Kitada 2016). Some scholars suggest that reparative renaming is critical to "regenerating" oppressive socio-cultural systems in ways that enable self-healing, greater public participation and assessments of positive and negative social impacts of memorialization on subaltern lives (Sheehan 2020).

The use of memorial toponymic practices for ethical aims opens up important avenues of analysis and casts great significance on the name protests in Paris and elsewhere. However, reparative place naming encounters limits and obstacles. Despite growth in commemorative challenges and changes, many public groups – even those sympathetic to marginalized communities – oppose giving up established place identifiers. Continuing inequalities and social wounds within communities severely limit subaltern memory-work. Reparative approaches depend upon a redistribution or sharing of naming power yet unrealized. In addition, toponymic resistance is not just carried out by oppressed populations and their allies but also deployed by dominant groups opposed to the removal and negative re-interpretation of existing memorial toponyms. These groups have at times been successful in misrepresenting reparative change as an erasure rather than a restoration of history.

There can also be different opinions among oppressed groups about how best to involve, if at all, place naming in a reparative politics (Brocket 2021). Heritage from below is a "force field" of different collective interests and intersectional identities that converge, diverge or mix in ways that defy binaries (Seaton 2001). Marginalized groups sometimes clash over "commemorative surrogation", the process of choosing whose name and memory is the most suitable signifier to fill the historical and emotional cavities created by the death, diaspora and trauma of oppression (Alderman 2015). Place naming can also be a source of division when seen as trivial and not clearly connected to larger material struggles for well-being and empowerment. *Osez le féminisme* faced criticism such as this in Paris from women who questioned the importance of challenging street signs relative to advocating for equal rights and pay (Gee 2015).

Some activists and scholars argue that the use of official place renaming to elevate public recognition of minority groups can actually work to deny power and self-determination to these groups. This is because the authority for designating place names remains solely within the jurisdiction of geographical naming boards and related institutions. The heritages and identities of subaltern groups, while more recognized publicly, become "domesticated", appropriated and put into the service of the same settler state responsible for histories of discrimination and violence (Rose-Redwood 2016). Indeed, some authorities use memorial toponymic change as a tool of appeasement and avoid the difficult conversations and practices necessary for full reparative memory-work (Brasher et al. 2017).

State-sanctioned name removal and renaming, although pursued by many oppressed groups, "is by no means the only way to bring new 'toponymic worlds' into being". Counter-memory making can also happen through the "performative force [of] everyday, embodied speech acts" that don't seek institutional permission to exist (Rose-Redwood 2016, p. 202). New toponymic worlds also arise from insurgent, highly publicized occupations of memorial space, such as that seen along Paris roadways in 2015. In this respect, what is dismissed as a street renaming "stunt" is more fairly recast as a legitimate way of opening up new ways of representing, sensing, materializing and debating the past.

Finally, realizing the reparative possibilities rather than limits of place name commemoration requires re-situating toponymic scholars, specifically those from privileged, non-subaltern backgrounds. We must do more than offer scholarly critiques and remain neutral onlookers of naming struggles. But that does not mean our ideas and presumed intellectual authority should supersede the knowledge, lived experiences and political desires of minority activists and scholars (Rose-Redwood 2016). To repair the future and not perpetuate systems of oppression requires a model of solidarity in which we create "workspaces" in our classrooms, research groups and lived communities to support toponymic activism and socially responsible commemoration (Alderman and Rose-Redwood 2020).

## 2.7. References

Adebanwi, W. (2012). Glocal naming and shaming: Toponymic (inter)national relations on Lagos and New York's streets. *African Affairs*, 111(445), 640–661.

Alderman, D.H. (2002a). School names as cultural arenas: The naming of US public schools after Martin Luther King, Jr. *Urban Geography*, 23(7), 601–626.

Alderman, D.H. (2002b). Street names as memorial arenas: The reputational politics of commemorating Martin Luther King in a Georgia county. *Historical Geography*, 30, 99–120.

Alderman, D.H. (2015). Naming streets, doing justice? Politics of remembering, forgetting, and finding surrogates for African American slave heritage. In *Geographical Names as Cultural Heritage*, Choo, S. (ed.). Hyung Hee University Press, Seoul.

Alderman, D.H. and Inwood, J. (2013). Street naming and the politics of belonging: Spatial injustices in the toponymic commemoration of Martin Luther King Jr. *Social & Cultural Geography*, 14(2), 211–233.

Alderman, D.H. and Reuben, R.R. (2020). The classroom as "toponymic workspace": Towards a critical pedagogy of campus place renaming. *Journal of Geography in Higher Education*, 44(1), 124–141.

Atencio, R. (2018). Haunting pasts and the politics of street naming in Nicaragua and Brazil. *City & Society*, 30(1).

Azaryahu, M. (1996). The power of commemorative street names. *Environment and Planning D: Society and Space*, 14(3), 311–330.

Azaryahu, M. (2011). The critical turn and beyond: The case of commemorative street naming. *ACME: An International E-journal for Critical Geographies*, 10(1), 28–33.

Azaryahu, M. and Foote, K.E. (2008). Historical space as narrative medium: On the configuration of spatial narratives of time at historical sites. *GeoJournal*, 73(3), 179–194.

Bigon, L. (2009). Urban planning, colonial doctrines and street naming in French Dakar and British Lagos, C. 1850–1930. *Urban History*, 36(3), 426–448.

Bigon, L. and Dahamshe, A. (2014). An anatomy of symbolic power: Israeli road-sign policy and the Palestinian minority. *Environment and Planning D: Society and Space*, 32(4), 606–621.

Bourillon, F. (2016). Dé-nommer et re-nommer la rue ou comment accompagner la transformation de Paris à la fin du XIX^e siècle. *Romantisme*, 1(171), 29–39.

Bouvier, J.C. (2007). Le langage de la toponymie urbaine. Approche méthodologique. *Onoma*, 42, 23–38.

Brasher, J.P., Alderman, D.H., Inwood, J.F. (2017). Applying critical race and memory studies to university place naming controversies: Toward a responsible landscape policy. *Papers in Applied Geography*, 3(3–4), 292–307.

Brasher, J.P., Alderman, D.H., Subanthore, A. (2020). Was Tulsa's Brady Street really renamed? Racial (in)justice, memory-work and the neoliberal politics of practicality. *Social & Cultural Geography*, 21(9), 1223–1244.

Brocket, T. (2021). Governmentality, counter-memory and the politics of street naming in Ramallah, Palestine. *Geopolitics*, 26(2), 541–563.

Capdepón, U. (2020). Challenging the symbolic representation of the Franco dictatorship: The street name controversy in Madrid. *History and Memory*, 32(1), 100–130.

Carter, P., Butler, D.L., Alderman, D.H. (2014). The house that story built: The place of slavery in plantation museum narratives. *The Professional Geographer*, 66(4), 547–557.

Crețan, R. and Matthews, P.W. (2016). Popular responses to city-text changes: Street naming and the politics of practicality in a post-socialist martyr city. *Area*, 48(1), 92–102.

Desbrosse, X. (2014). Héros de l'Ouest, héros de l'Est : Toponymie et Guerre froide de la Libération à nos jours. In *La Guerre froide vue d'en bas*, Buton, P., Büttner, O., Hastings, M. (eds). CNRS Editions, Paris.

Doss, E. (2010). *Memorial Mania: Public Feeling in America*. University of Chicago Press, Chicago, IL.

Duminy, J. (2014). Street renaming, symbolic capital, and resistance in Durban, South Africa. *Environment and Planning D: Society and Space*, 32(2), 310–328.

Dwyer, O.J. and Alderman, D.H. (2008). Memorial landscapes: Analytic questions and metaphors. *GeoJournal*, 73(3), 165–178.

Ebron, P.A. (2014). Slavery and transnational memory: The making of new publics. In *Transnational Memory: Circulation, Articulation, Scales*, De Cesari, C. and Rigney, A. (eds). De Gruyter, Berlin.

Ferguson, G. (2019). The perceptions and effects of schools' names on black professional educators and their students. Unpublished Dissertation, Marshall University, Huntington, WV.

Forrest, C. (2018). What's in a name? A feminist reflection on street name changes in Durban. *Agenda*, 32(2), 53–61.

Gee, O. (2015). "Sexist" Paris streets renamed in feminist stunt. *The Local*, 25 August [Online]. Available at: https://www.thelocal.fr/20150826/paris-neighbourhood-gets-a-feminist-makeover.

Giraut, F. and Houssay-Holzschuch, M. (2016). Place naming as dispositif: Toward a theoretical framework. *Geopolitics*, 21(1), 1–21.

Giraut, F., Guyot, S., Houssay-Holzschuch, M. (2008). The name game: Toponymic changes in South Africa. *L'Espace géographique*, 37(2), 131–150.

Johnson, K. (2019). Confederate named high schools: Some say no more, others say changing names unnecessary. *Montgomery Advertiser,* 3 June [Online]. Available at: https://www.montgomeryadvertiser.com/story/news/education/2019/06/03/jefferson-davis-birthday-alabama-holiday-why-three-montgomery-high-schools-named-confederates/2638908002/

Jones, O. and Garde-Hansen, J. (eds) (2012). *Geography and Memory: Explorations in Identity, Place and Becoming*. Springer, London.

Karimi, A. (2016). Street fights: The commodification of place names in post-Taliban Kabul City. *Annals of the American Association of Geographers*, 106(3), 738–753.

Kearns, R.A. and Berg, L.D. (2002). Proclaiming place: Towards a geography of place name pronunciation. *Social & Cultural Geography*, 3(3), 283–302.

Kearns, R.A. and Lewis, N. (2019). City renaming as brand promotion: Exploring neoliberal projects and community resistance in New Zealand. *Urban Geography*, 40(6), 870–887.

Kitada, E. (2016). Commemorating racial violence: Street naming and segregation in New York City, 1999. *Nanzan Review of American Studies*, 38, 21–34.

Light, D. and Young, C. (2014). Habit, memory, and the persistence of socialist-era street names in postsocialist Bucharest, Romania. *Annals of the Association of American Geographers*, 104(3), 668–685.

Madden, D.J. (2018). Pushed off the map: Toponymy and the politics of place in New York City. *Urban Studies*, 55(8), 1599–1614.

Mamvura, Z., Muwati, I., Mutasa, D.E. (2018). Toponymic commemoration is not for one sex: The gender politics of place renaming in Harare. *African Identities*, 16(4), 429–443.

Masalha, N. (2015). Settler-colonialism, memoricide and indigenous toponymic memory: The appropriation of Palestinian place names by the Israeli state. *Journal of Holy Land and Palestine Studies*, 14(1), 3–57.

Medway, D., Warnaby, G., Gillooly, L., Millington, S. (2019). Scalar tensions in urban toponymic inscription: The corporate (re)naming of football stadia. *Urban Geography*, 40(6), 784–804.

Mitchelson, M.L., Alderman, D.H., Popke, E.J. (2007). Branded: The economic geographies of streets named in honor of Reverend Dr. Martin Luther King, Jr. *Social Science Quarterly*, 88(1), 120–145.

Murphyao, A. and Black, K. (2015). Unsettling settler belonging: (Re)naming and territory making in the Pacific Northwest. *American Review of Canadian Studies*, 45(3), 315–331.

Muzaini, H. (2015). On the matter of forgetting and memory returns. *Transactions of the Institute of British Geographers*, 40(1), 102–112.

Naef, P. (2019). Memorial entrepreneurs in post-conflict tourism. In *Tourism and Hospitality in Conflict Ridden Destinations*, Isaac, R.K., Çakmak, E., Butler, R. (eds). Routledge, London.

Nora, P. (1989). Between memory and history : les lieux de mémoire. *Representations*, 26, 7–24.

Poon, L. (2015). Mapping the sexism of city street names. *CityLab*, 4 November [Online]. Available at: https://www.citylab.com/equity/2015/11/mapping-the-sexism-of-city-street-names/414094/.

Rose-Redwood, R.S. (2008). From number to name: Symbolic capital, places of memory and the politics of street renaming in New York City. *Social & Cultural Geography*, 9(4), 431–452.

Rose-Redwood, R.S. (2014). Sixth avenue is now a memory: Regimes of spatial inscription and the performative limits of the official city-text. *Performativity, Politics, and the Production of Social Space*, Glass, M.R. and Rose-Redwood, R. (eds). Routledge, New York.

Rose-Redwood, R.S. (2016). Reclaim, rename, reoccupy: Decolonizing place and the reclaiming of PKOLS. *ACME: An International E-Journal for Critical Geographies*, 15(1), 187–206.

Rose-Redwood, R.S., Alderman, D., Azaryahu, M. (eds) (2018). *The Political Life of Urban Streetscapes: Naming, Politics, and Place*. Taylor & Francis, London.

Rose-Redwood, R.S., Sotoudehnia, M., Tretter, E. (2019). Turn your brand into a destination: Toponymic commodification and the branding of place in Dubai and Winnipeg. *Urban Geography*, 40(6), 846–869.

Ryan, M.L., Foote, K., Azaryahu, M. (2016). *Narrating Space/Spatializing Narrative: Where Narrative Theory and Geography Meet*. Ohio State University Press, Columbus, OH.

Seaton, A.V. (2001). Sources of slavery-destinations of slavery: The silences and disclosures of slavery heritage in the UK and US. *International Journal of Hospitality & Tourism Administration*, 2(3–4), 107–129.

Sheehan, R. (2020). Rethinking memorial public spaces as regenerative through a dynamic landscape assessment plan approach. *Regenerative Urban Development, Climate Change & the Common Good*, Caniglia, B., Frank, B., Knott Jr., K., Sagendorf, K.S., Wilkerson, E.A. (eds). Routledge, London.

Spocter, M. (2018). A toponymic investigation of South African gated communities. *South African Geographical Journal*, 100(3), 326–348.

Sumartojo, S. (2016). Commemorative atmospheres: Memorial sites, collective events and the experience of national identity. *Transactions of the Institute of British Geographers*, 41(4), 541–553.

Swart, M. (2008). Name changes as symbolic reparation after transition: The examples of Germany and South Africa. *German Law Journal*, 9(2), 105–121.

Till, K.E. (2012). Wounded cities: Memory-work and a place-based ethics of care. *Political Geography*, 31(1), 3–14.

Tolia-Kelly, D.P., Waterton, E., Watson, S. (eds) (2016). *Heritage, Affect and Emotion: Politics, Practices and Infrastructures*. Routledge, London.

Tunbridge, J.E. and Ashworth, G.J. (1996). *Dissonant Heritage: The Management of the Past as a Resource in Conflict*. John Wiley & Sons, Hoboken, NJ.

Vuolteenaho, J., Lappalainen, H., Ainiala, T. (2019). A slightly slummier area? Negotiations of place-bound identities through social spatializations and unofficial toponyms. *Journal of Sociolinguistics*, 23(2), 140–162.

Wanjiru, M.W. and Matsubara, K. (2017). Street toponymy and the decolonisation of the urban landscape in post-colonial Nairobi. *Journal of Cultural Geography*, 34(1), 1–23.

Ward, G. (2015). The slow violence of state organized race crime. *Theoretical Criminology*, 19(3), 299–314.

Wideman, T.J. and Masuda, J.R. (2018). Assembling "Japantown"? A critical toponymy of urban dispossession in Vancouver, Canada. *Urban Geography*, 39(4), 493–518.

Wilce, J. (2012). The grammar of politics and the politics of grammar: From Bangladesh to the United States. In *A Cultural Approach to Interpersonal Communication: Essential Readings*, Monogham, L., Goodman, J.E., Robinson, J.M. (eds). Blackwell Publishing, Chichester.

# 3

# The Named, Lived and Contested Environment: Towards a Political Ecology of Toponymy

**Sébastien BOILLAT**

*University of Bern and Bern University of Applied Sciences and University of Neuchâtel, Switzerland*

This contribution aims to lay the foundations for a field of research that considers the role of place names both in the expression of the human–environment relationship and in the power issues related to this relationship. This field of research, called the political ecology of toponymy, extends the study of place names to include processes of naming based on critical toponymy, as well as the use and mobilization of these names by the social groups involved. It also focuses on microtoponyms that express a set of relationships between individuals of a social group and environmental elements. The power issues concerning microtoponyms allow us to shed light on broader issues concerning environmental perceptions and related ethics defended by these groups.

## 3.1. Introduction

"Downstream, it is called *Pintu Mayu*, but the real name of this river is the *Río Colón*..." one of the indigenous leaders of an agrarian union near Cochabamba (Bolivia) insists. But why claim a *colonial* toponymy against the one based on the term *mayu*, which means river in Quechua language? Rejecting the name of *Colón*

*The Politics of Place Naming*,
coordinated by Frédéric GIRAUT and Myriam HOUSSAY-HOLZSCHUCH. 

would mean ignoring the fact that *Colón* is also the name of the agricultural zone that borders the river, and that appears on the collective title of the agrarian syndicate, obtained thanks to the agrarian reform of 1953. A title that, together with a rough map drawn by a topographer of the time, guarantees not only the rights of access to the land, but also to a part of the water of the said river to the members of the syndicate.

This anecdote, collected in Bolivia (Boillat 2014), highlights the overlapping political, institutional, relational and environmental characters of the toponym. In this case, the colonial nature of a place name loses relevance against its importance as a *dispositif* for legitimizing access rights, and as an expression of an intimate relationship between the members of a community and a specific natural element. In other words, the influence of the signifier is lightened in the face of the signified of a complex human–environment relationship charged with access rights, management rules and individual and collective experiences.

Such a perspective goes beyond the study of names as such to include processes such as naming, as well as the use of place names. I argue that it opens up new horizons for toponymy as a multidisciplinary field of research. Although some multidisciplinary uses of toponymy, such as that of a substitute for archeology, are considered by many to be obsolete (Zadora-Rio 2001), other interdisciplinary approaches involving toponymy have emerged. These approaches link toponymy with landscape ecology (Derungs and Penko Seidl 2008; Purves 2014 and Villette and Purves 2018), cultural and linguistic heritage (Basso 1988; Cruikshank 1990), environmental management and ethics (Boillat et al. 2013; Samakov and Berkes 2017) and political geography (Berg and Vuolteenaho 2009; Rose-Redwood et al. 2010 and Giraut and Houssay-Holzschuch 2016).

At first glance, these approaches seem anecdotal and disconnected from each other, drawing on the toponymic repertoire according to their needs. Should toponymy be reduced to a database, or even a research approach instrumentalized by other disciplines? On the contrary, toponymy can become a bridge, a link between these different disciplines, consistent enough to raise new research questions. This chapter supports this position. In particular, I will attempt to connect these new developments to provide evidence of the role of toponymy as a cornerstone for understanding human–environment relations.

Within this framework, I argue that two main fields of research deserve closer interaction. The first is the one that links toponymy with the environment and focuses on the relationships between place names and language with environmental conceptualization, management and ethics. It allows us to understand how a social group conceptualizes, represents, values and transforms a landscape in thought and language (Mark et al. 2011). This process plays a fundamental role in human–environment management and interaction. The second is critical toponymy (Berg and Vuolteenaho 2009; Rose-Redwood et al. 2010 and Giraut and Houssay-Holzschuch 2016), which shows that toponymy is an expression of power issues. These two fields of research have in common the fact that they are interested not only in the toponymic corpus and its signifiers and signifieds from a linguistic and historical perspective, but also in the contested processes of naming, of the use of place names and of their relation to elements in space. They therefore go beyond the issues related to toponymic meanings alone.

The interaction of these two fields of research leads us to consider toponymy as both an *ecology*, in the sense of a conceptualization of environmental elements endowed with their own dynamics and based on a very real human–environment relationship, and a *political* issue linked to the exercise of power through the transformation of space into entities loaded with cognitive elements. This reflection, built on theoretical and empirical bases, leads to an argument in favor of a *political ecology of toponymy* that deals with the relations between the processes of naming, onomastics and the use of place names. This chapter aims to establish the foundations of this field of research and to explore its issues.

Political ecology aims to highlight the power issues in society–environment relations and pays particular attention to the way environmental problems are formulated and how these formulations respond to different interests (Robbins 2009). My main point of entry is that a political ecology approach to toponymy allows us to explore the conceptualization, production and use of spatial entities – a spatial ontology – as an act of power that is accumulated, rethought and modified over time. These processes create winners and losers and also significantly shape the relationship between individuals, social groups and elements of space and landscape.

This chapter is structured as follows: firstly, I review the various existing approaches that mobilize toponymy in the understanding of the human–environment relationship in a broad sense. Based on this state of the art, I then propose an analytical framework that identifies the key dimensions of a political ecology approach to toponymy.

## 3.2. The decline of toponymy as a substitute for archeology

Until the 1980s, a common application of toponymy was its use to reconstruct the history of land use and geographic space. This approach, linked to regional geography and settlement archaeology, claimed to identify the origin of inhabited sites to an ethnic group and/or to economic, religious and political activities by the signifier of its name. It was based on classical onomastics focused on linguistics and on the origin of place names. For example, Chambon (1999), quoted by Zadora-Rio (2001), attempts to reconstruct the settlement dynamics of the Carolingian period by analyzing the names of villages and localities in the Issarlès region and their relationship with the establishment of religious sites and cultivation areas.

This approach is strongly criticized by Zadora-Rio (2001) in light of new archaeological discoveries whose density often exceeds the toponymic grid. Moreover, the dating of many sites rarely corresponds to the period attributed to the origin of place names. There is thus a spatial and temporal mismatch between archaeological sites and place names. This discrepancy, Zadora-Rio (2001) argues, is related to two wrong assumptions. First, place names do not always correspond to the features and uses of the landscape that their signifiers suggest. Second, place names do not remain fixed after a hypothetical era of settlement idealized as foundational and definitive. On the contrary, the naming system is often disconnected to the materiality of the habitat and the elements of the landscape. The localities are not a set of superimposed layers but a constantly changing system in which the ancient elements are permanently updated, recomposed and transformed (Zadora-Rio 2001).

Given this evidence, toponymy would lose its use as a key to interpreting the landscape and its occupation. However, many recent studies question the complete separation between place names and landscape features. These approaches clearly move away from a purely archaeological and historical perspective to incorporate contemporary aspects of the significance and use of place names. First, there are studies that seek to explore contemporary and historical vernacular forms of landscape categorization. While some of these studies explicitly exclude place names as a source of information (Hunn and Meilleur 2010), others refer to them by triangulation with other approaches such as ethnography, participatory mapping and the extraction of geographic databases (Derungs and Purves 2014; Wartmann et al. 2016). Second, another field of research focuses on extending the spatial resolution of toponymy by integrating the study of proper names of landscape features at a more detailed scale (Jones 2016; Villette and Purves 2018). This "parcel-based" toponymy or micro-toponymy could make it possible to reconcile the spatiality of occupied sites and localities and constitute a corpus closer to the materiality of the habitat and the landscape. Third, the naming of a place, its use and its interpretation can be studied as a key social phenomenon that shapes the relationship between humans and geographical space. To harness its full potential, the study of place names must therefore extend its scope beyond classical onomastics and include the processes of naming, interpreting and using place names.

## 3.3. Toponymy and ecology: another divorce, another reconciliation

The current focus of landscape ethnoecology sheds light on the place given to toponymy in the study of human–environment relations. Landscape ethnoecology builds on the discipline of landscape ecology (Naveh and Lieberman 2013), which seeks to describe and categorize a set of ecosystems and understand the ecological and socio-ecological processes that link them. Landscape ethnoecology specifically goes beyond the study of the environment by experts to focus on how social groups perceive and categorize landscape elements (Hunn and Meilleur 2010).

Landscape ethnoecology is intended to be a participatory approach that aims not only to enhance vernacular, indigenous and citizen knowledge, but also to democratize environmental management (Johnson and Hunn 2010). It is part of a paradigm shift in ecological science that moves away from a static, mechanistic view of ecosystems to emphasize their complex, nonlinear, unpredictable and adaptive properties (Gunderson 2001). These complex properties justify the consideration of holistic knowledge and in particular the valuation of traditional and local ecological knowledge (Berkes et al. 2000). In the specific case of ecosystem classification, the recognition of the ecosystem as physically open (Jorgensen et al. 2007) paves the way to the use and valuation of vernacular landscape classifications (Berkes et al. 1998; Johnson and Hunn 2010).

However, initially, landscape ethnoecology explicitly rejects the use of place names (Hunn and Meilleur 2010). Toponymy is seen as an habitat naming system that remains separated from landscape categorization, as Zadora-Rio (2001) sustains. The approach favored by landscape ethnoecology is ethnographic and focuses on elucidating local categories of landscape units, often based on vegetation, in the local linguistic corpus (Davidson-Hunt and Berkes 2003; Mark et al. 2011). More recently, the ethnographic approach has expanded to include the combination and comparison of different sources. For example, the combination of linguistic studies and participatory mapping shows that local classifications are not limited to vegetation but include important topographic and hydrographic features (Wartmann, Purves 2017; Wartmann and Purves 2018).

Other studies seek to reconstruct vernacular characterizations of the landscape (called spatial folksonomies) based on written sources by automating the interpretation of large databases (Derungs, Purves 2016; Wartmann et al. 2016). In this context, toponymy first reappears as a link between localization and categorization. On the one hand, toponymy is used to georeference descriptive information of landscape elements taken from different historical or recent databases, and, on the other hand, it is used to compare these databases with each other and with expert classifications (Derungs and Purves 2014, 2016). As Wartmann et al. (2016) indicate, this approach draws a clear separation between the bodies of knowledge for differentiating space into places formulated by Agnew (2014): toponymy-based localization (location), categorization-based description of landscape elements (locale) and emotional attachment to a place (sense of place).

Even more recently, studies are crossing this barrier to consider toponymy as a possible source of information about landscape features and classifications. For example, a large toponymic database allows for the recognition of dominant categorizations in a given language area (Villette and Purves 2018). On the other hand, historical toponymic data allows us to understand the environmental history and risks associated with portions of the territory at a very fine scale (Jones 2016). These studies are systematically based on microtoponymy, a body of oral or written local knowledge designating portions of space on a very fine scale such as fields, pastures, hamlets, farms, streams and wooded areas (Noël 1995; Boillat 2018).

It is therefore the shift of scale from classical toponymy to microtoponymy that makes this connection possible. Studies conducted in different cultural settings (Taller de Tradición Oral and Beaucage 1996; Penko Seidl 2008; Boillat et al. 2013 and Villette and Purves 2018) converge on several shared characteristics of microtoponyms. Firstly, microtoponyms are not restricted to the built environment, but extend to fields, forests, rivers and pastures, characterized by an interaction between humans and the environment, in other words, "nature-society hybrids" (Zimmerer 2000). Second, they cannot be *resolved* (unambiguously localized) because they are repetitive on a larger scale despite a local uniqueness (Taller de Tradición Oral and Beaucage 1996; Boillat et al. 2013; Villette and Purves 2018).

Third, a large portion of the microtoponymic corpus consists of signifying lexemes. The fact that these lexemes follow Zipf's law (Villette and Purves 2018), that is, their frequency decreases on a logarithmic scale, as observed in everyday language (Zipf 1935), shows their greater affinity with a linguistic body of common nouns than with a body of proper nouns. These particularities mean that microtoponyms cannot be clearly attributed to a corpus of localization (location) or to a corpus of categorization of landscape elements (locale) differentiated by Agnew (2014). Nor can it be completely separated from the sense of place, as I will show in the next section.

## 3.4. From cultural heritage to environmental ethics: indigenous place names and beyond

I now consider another approach of toponymy that emphasizes the cultural value of the toponymic corpus. In Europe, this approach considers the oral forms of microtoponymy as an endangered heritage related to small rural communities (Crivelli 2013). On the American continent, it is more closely linked to the indigenous context and is part of an ethnographic field tradition inaugurated by Franz Boas (Boas 1934) and represented by numerous studies (Basso 1988; Cruikshank 1990; Collignon 1996; Fair 1997; Heikkilä 2014). These studies emphasize the cultural importance of place names for the indigenous peoples involved, as links to collective memory, to the anchoring and transmission of indigenous knowledge, to the attachment to the territory and to the role of the place in a specific worldview.

Although it belongs to the discipline of social anthropology and shares with landscape ethnoecology its predilection for indigenous contexts and its attachment to cultural diversity, the ethnographic approach differs from it in that it integrates several elements that clearly go beyond the separation between location, locale and sense of place mentioned above. In this framework, toponymy is no longer simply a source of spatial or lexical data subordinated to a classification of the landscape. It is above all interpreted as a body of knowledge of its own and of crucial importance for the people and communities that hold it, from a material point of view as well as from the point of view of identity, culture and spirituality.

From a material point of view, toponymy is crucial for locating and characterizing resources in often extensive activities such as mountain agriculture, pastoralism, hunting, foraging or fishing. It is also crucial for the management of these resources when they are shared by a social group. As I argued in a previous publication (Boillat 2018), microtoponymy can be seen as a vernacular micro-territorial network that allows for the management of common territorialized resources such as pastures, water sources, collective crop rotations, hunting, fishing and foraging areas. It does provide information on the location and state of the resource, but it also makes it possible to anchor and guarantee customary use rights and to defend them both internally and externally. In this sense, toponymy plays a significant role in the implementation of institutions for managing the commons (Ostrom 1990). However, apart from one study showing the links between management of the commons and sacred places (Samakov and Berkes 2017), the field of environmental commons governance has not yet given attention to toponymy.

To limit the interpretation of place names to the location of natural resources and their management is, however, subject to the critique of functionalism. Notions of natural resources or "cultural ecosystem services" assume an instrumental and market-based view of the environment that originates in modern economics. However, ethnographic studies that give indigenous and local partners a fair say have shown that toponyms may express a relationship with spatial elements that go beyond materiality (Fair 1997). This relationship becomes a fundamental component of individual, social and cultural identity (Basso 1988); it connects with a sense of place in a historical, cosmological and sacred perspective (Bradley and Kearney 2009; Heikkilä 2014). More than containers of resources to be exploited, named places become relational spaces where identity, attachment, religiosity, and subsistence and economic activities intertwine.

From such a perspective, it is fundamental to conceptualize the named place not as an object of knowledge, but as a specific ontological category. In the literature from indigenous contexts, there are numerous references to the phenomenon of place personification. Among the Inupiat of Alaska, there are places that "speak" (Fair 1997), and among the Apache of Arizona, people who speak to places (Basso 1988). In the Andes, places "get angry" and people direct offerings to them (Martínez 1989; Spedding 1992 and Boillat et al. 2013) or consider them as kin (Salas Carreño 2016). In such context, naming a place is like naming a person, and naming it incorrectly is disrespectful, hence the need to "know the right words" (Cruikshank 1990).

This personification should not be interpreted as a naive view of animism, which would be the assignment of life and spirit to inert objects (Boillat 2020). On the contrary, this personification is closer to an animic worldview, in which the very concept of life and thought must be reinterpreted (Ingold 2006). In this vision, it is the idea of movement, change and relation that precedes the differentiation of the entities in objects (Ingold 2006). The very concept of life being that of transformation and relationship, in which animals, plants, landscapes and inhabited places are constantly transforming in a dynamic interaction with human and spiritual beings (Ingold 2011).

These reflections that consider cultural diversity in the notion of place, as a named but sometimes revealed entity, open up new perspectives in the human–environment relationship. The environment can neither be seen as a simple resource, nor as a pure product of human imagination or "social construction"; it is the very notion of separation between nature and culture that is questioned (Descola 2006). In this framework, the environment is no longer a nature with its intrinsic value separated from the human, but a set of relational events. Joining similar conclusions based on epistemology (Latour 1999) and environmental ethics (Leopold 1949; Plumwood 2002), this position clearly goes beyond the indigenous context. It becomes a questioning of our own modern societies in the light of cultural diversity, the true goal of anthropology according to Descola (2006).

The role of place in this vision of the environment becomes crucial: the undifferentiated environment loses its meaning and it is the revealed, named and experienced place that becomes the anchor of a relationship with the elements constituting that environment. In other words, (micro)-toponyms become dynamic, unpredictable, unique and irreplaceable "relational places" or "place-ecosystems" (Boillat et al. 2013). These relational nodes connect humans with natural phenomena and also play a crucial role in human-to-human relationships. The animic perspective of place must therefore be complemented by a science of human–human–environment relations (Gottschlich et al. 2017) which integrates the dimension of power. To do this, I propose to complement the perspective of lived place with that of disputed place.

## 3.5. The disputed toponymy: critical perspectives

Critical toponymy represents the political perspective of toponymy, which links toponymy with political geography. As this perspective is extensively detailed in this volume, the aim of this section is to understand its implications for an approach of the "place-ecosystem" that is sensitive to questions of power, and in this way to lay the foundations for a political ecology of toponymy.

First, critical toponymy adds the study of place naming processes to the study of place names. Studying place naming processes makes power relations and issues visible. The action of naming is associated with a dispositif in the sense of Foucault, who interprets the exercise of power through the combination of discourses, regulations, material goods and actors' strategies (Giraut and Houssay-Holzschuch 2016). It is also similar to a form of governmentality that links knowledge and power. Critical toponymy proposes an analytical framework centered on three main dimensions: the context, the technology of nomination and the actors involved.

Context is given as 1) a conquest, or forcible subjugation of a territory where new place names are imposed, 2) a revolution, a radical change in political order that motivates name changes, 3) an emergence, which follows developments or differentiations that call for the creation of new place names and 4) a commodification, where the creation of place names follows an appropriation of common goods by private actors. The technologies used include 1) "cleansing", which seeks to erase and replace place names that evoke a culture, language or political order deemed undesirable, 2) foundation, which inscribes cultural and political references of a new order into place names, 3) restoration, which seeks to restore past place names in a reparative manner and 4) promotion, which seeks to create value, often market value, from a name. The actors, finally, are distinguished between the state, private actors and civil society (Giraut and Houssay-Holzschuch 2016).

Until now, research in critical toponymy has focused on either large-scale or on urban spaces, with little discussion of the more specific context of microtoponyms and its implications for the human–environment relationship. In a previous contribution (Boillat 2018), I attempted to outline some possible applications of critical toponymy to this type of toponym, addressing two main points that lay the groundwork to build an analytical framework to make the political ecology of toponymy operational.

First, the relationship between microtoponymy, landscape categorization and environmental management brings us back to the more general question of who manages the environment, in the sense of environmental decision-making. It consists of discerning how different claims and land use practices such as agriculture, pastoralism or nature conservation are articulated through the reference and use of specific place names. In this context, existing microtoponyms can be either contested, ignored in favor of other divisions such as vegetation units, or over-interpreted as a reliable source of ecological descriptions. It is mainly a struggle involving one social group, often local, with another group claiming similar or different uses of the land, or with actors operating at a larger scale.

In this case, the issue at stake is not the signifiers of microtoponyms as such, but their possession, their valuation and the scope of their territorial resolution. Local communities, sometimes in competition, will seek to legitimize their rights of use over a territory by controlling the microtoponymic corpus. This is done either by making part or all of this corpus visible, or by making it secret through its sacredness. In the latter case, external actors may choose to "trample", that is, ignore this invisible repertoire in order to delegitimize the rights of the communities, or to "plunder" it, which consists of reappropriating the corpus by constructing databases that would escape local control and supplant its legitimacy (Boillat 2018).

To go further, the valuation or devaluation of microtoponyms or of some of them is associated with the promotion of a certain vision of the human–environment relationship. The division of a territory into vegetation units, into sometimes sacred places or into GIS raster cells, tells a lot about the power exercised and about the legitimization of a certain form of territorial and ecological knowledge. When associated with a vision of place that plays a key role in the construction of a cultural and relational identity, the recognition and valuation of microtoponyms can constitute an alternative to forms of power inherited from colonization (Boillat 2020). In this sense, it constitutes a demand for cognitive justice, understood as taking into account the diversity of knowledge systems through counter-hegemonic actions (Visvanathan 2009).

Secondly, microtoponyms also represent power dispositifs – or more precisely, *micro-dispositives* – within the social groups that share and use them. This highlights the importance of nomination processes again. Although these processes often remain obscure because they are difficult to situate in their historical context, the process of emergence linked to the development of activities such as the establishment of fields, pastures or the founding of villages seems to dominate (Boillat 2018). In that previous study (Boillat 2018), I have shown that in the Andean case, these processes of domination are often linked to official or local religious institutions with a preponderance of founding myths. In this context, it becomes difficult to clearly identify the process of naming in relation to the use of the microtoponym in oral or written form, or in negotiations of usage rights or during religious incantations.

## 3.6. Towards a political ecology of toponymy

To integrate toponymy into a political ecology perspective, I argue that toponymy must not only integrate the processes of naming, but also those of use. In a context that involves microtoponymy as a human–environment relationship and where orality dominates, these two processes are closely intertwined: there is both uncertainty as to the attribution of the name and to the risk of forgetting a toponym fallen into disuse. The power issues surrounding this intertwinement may be internal or external to the social group, or *microtoponymic community*, that shares knowledge of these place names. The following examples, taken from ethnographic data from the Bolivian Andes, can illustrate these processes.

Internal issues: the named place *Qolqe Yupana*, in Quechua "where the money is counted", is located on the path that small farmers living in scattered mountain dwellings use to return from the market. People have become used to sitting on a rock to count the money earned, some sharing the belief that the money counted in this place will reach higher sums than if it is counted elsewhere. Place naming occurs through the emergence of an economic activity which, however, implies a certain visibility of the money earned among villagers and the guarantee of rights of passage over this place. In this case, the use of the toponym acquires a material dimension, that of following the road, a social dimension, that of the rights of passage and village solidarity, and spiritual dimension, the alleged power to "multiply the money" associated with this place.

External issues: the named place *Colón* is on the title deed of an agrarian syndicate in the peri-urban area of Cochabamba. The syndicate has been in conflict with the authorities of the Tunari National Park, which has put the syndicate's lands under protection (Boillat et al. 2008). The place is associated with the river known as *Río Colón,* after the old Colón brewery that exploited the water of the river before its closure. The name of *Colón* gives historical legitimacy to the rights of access to both land and water to the syndicate. However, it is the name *Pintu Mayu* (literally "Pinto's River") given to the same river downstream after a former landlord, which appears on the park's development plans, thereby contributing to the invisibility of the syndicate's land rights. In this case, the power issue concerns both the naming and the use of the toponym, highlighting the strategies of visibilization and invisibilization of the different actors.

These examples and these theoretical considerations make it necessary to extend the framework of critical toponymy to a broader framework that lays the foundation for a political ecology of toponymy. These extensions go mainly in two directions (Figure 3.1). First, the notion of naming is coupled with that of usage, as these two processes are often strongly intertwined. These usages are themselves interpreted as an expression of power that is internal and external to a group sharing the toponymic repertoire.

From an "internal" perspective, the issue is mainly the legitimization of a privileged relationship between individuals, families, clans and groups within a community that uses a shared territory. The question of who controls which part of the toponymic body and who is able to influence it, is key. This control takes a material form, through the concretization of de facto uses, an institutional form, through the legitimization of these uses by rights of access (oral or written, formal or informal), and a spiritual form, through the legitimization of a relationship referring to entities of a supernatural, ancestral-historical and identity-based order.

In an "external" perspective of struggles between actors, where one actor seeks to defend their access rights against another actor, the issue is the recognition of the microtoponymic body. In this case, it is often a large part or the entire corpus of microtoponymy that is at stake, with strategies ranging from *visibility* for the "defending" actor, that is, the display of microtoponyms by the holders of this corpus, to *secrecy* or concealment, and for the "attacking" actor, *trampling*, ignoring and devaluing this corpus, or *plundering* it through appropriation.

Second, the consideration of an environmentality related to microtoponyms appears to be crucial. This implies interpreting microtoponyms as bridges between location, locale and sense of place, and highlighting the ontological positions linked with the claims of specific microtoponymic bodies.

## 3.7. Conclusion

In this chapter, I have attempted to show the evolving role of toponymy in understanding the human–environment relationship in a broad sense. In line with the basic premise of political ecology, this relationship is a crucial power issue that defines and questions the relationships between individuals, communities and environmental elements, in this case more precisely the elements of the local landscape. Toponymy, as an entry point for the perception of environmental elements in the territory, crystallizes this relationship and can be understood from a political ecology perspective.

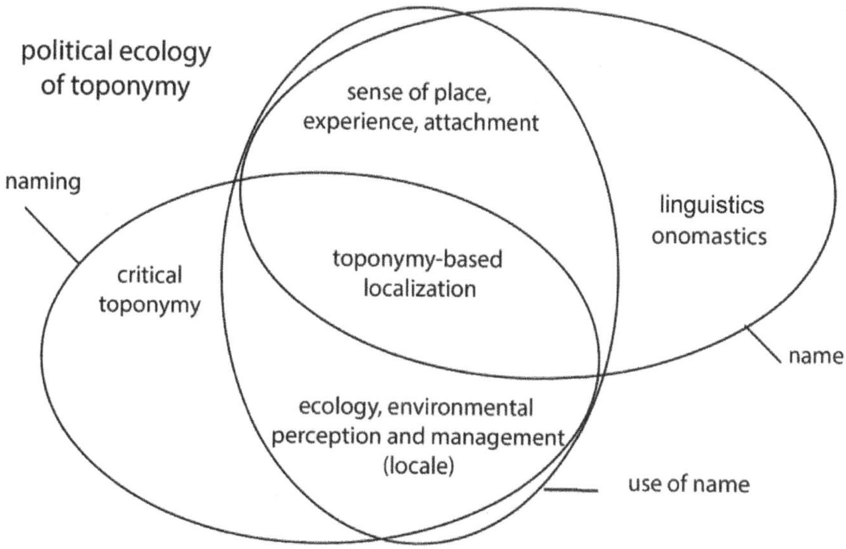


**Figure 3.1.** *Analytical framework for a political ecology of toponymy*

To implement such an approach, two elements are key. First, we need to go beyond the study of toponyms as signifiers, the name as such as seen by classical onomastics. We need to extend toponymy needs to the processes of naming, as advocated by critical toponymy, as well as to the processes of use and mobilization of place names. Both naming and usage are social phenomena that involve winners and losers and are the object of power struggles within and between social groups.

Second, the integration of microtoponyms, which act as a bridge between location, description and place attachment, enables linking toponymy and environmental features. These microtoponyms are often unique only at a local scale and do not allow for toponymic resolution at a larger scale. Often very dense (Kingston 2009), they represent both a network of human–environment relationships and the expression of a strong local identity and heritage. They can therefore play a key role in the recognition and development of an inclusive and emancipatory form of environmental ethics and management.

## 3.8. References

Agnew, J.A. (2014). *Place and Politics: The Geographical Mediation of State and Society*. Routledge, London.

Basso, K.H. (1988). Speaking with names: Language and landscape among the Western Apache. *Cultural Anthropology*, 3(2), 99–130.

Berg, L.D. and Vuolteenaho, J. (eds) (2009). *Critical Toponymies: The Contested Politics of Place Naming*. Ashgate, Aldershot.

Berkes, F., Kislalioglu, M., Folke, C., Gadgil, M. (1998). Exploring the basic ecological unit: Ecosystem-like concepts in traditional societies. *Ecosystems*, 1(5), 409–415.

Berkes, F., Colding, J., Folke, C. (2000). Rediscovery of traditional ecological knowledge as adaptive management. *Ecological Applications*, 10(5), 1251–1262.

Boas, F. (1934). *Geographical Names of the Kwakiutl Indians*. Columbia University Press, New York City.

Boillat, S. (2014). *Protective Mountains, Angry Lakes and Shifting Fields: Traditional Ecological Knowledge and Ecosystem Diversity in the Bolivian Andes*. Scholars' Press, Newcastle upon Tyne.

Boillat, S. (2018). Le rôle de la microtoponymie dans la reconnaissance des savoirs autochtones en gestion territoriale rurale : enjeux et défis. In *Hommes-Milieux. Croisement des savoirs et méthodologie de l'interdisciplinarit*, Pomade, A. (ed.). Presses universitaires de Rennes.

Boillat, S. (2020). Decolonizing ecological knowledge: Transdisciplinary ecology, place making and cognitive justice in the Andes. In *The Elgar Companion to Geography, Transdisciplinarity and Sustainability*, Sarmiento, F. and Frolich, L. (eds). Edward Elgar Publishing, Cheltenham.

Boillat, S., Rist, S., Serrano, E., Ponce, D., Delgadillo, J. (2008). Struggling "ontological communities": Transformations of conservationists' and peasants' discourses in the Tunari National Park in Bolivia. In *Geographica Bernensia*, Frey, M. and Haller, T. (eds). Bern, 37–80.

Boillat, S., Serrano, E., Rist, S., Berkes, F. (2013). The importance of place names in the search for ecosystem-like concepts in indigenous societies: An example from the Bolivian Andes. *Environmental Management*, 51(3), 663–678.

Bradley, J. and Kearney, A. (2009). Manankurra: What's in a name? Placenames and emotional geographies. In *Aboriginal Placenames: Naming and Re-Naming the Australian Landscape*, Hercus, L. and Koch, H. (eds). Australian National University, Canberra.

Chambon, J.-P. (1999). L'agencement spatial et fonctionnel des vicairies carolingiennes dans le midi de la Gaule : une approche linguistique. *Revue de Linguistique Romane*, 63, 55–173.

Collignon, B. (1996). *Les Inuits, ce qu'ils savent du territoire*. L'Harmattan, Paris.

Crivelli, R. (2013). Toponymie : un héritage précieux. *Revue de géographie alpine* [Online]. Available at: http://journals.openedition.org/rga/1873 [Accessed 25 August 2022].

Cruikshank, J. (1990). Getting the words right: Perspectives on naming and places in Athapaskan oral history. *Arctic Anthropology*, 27(1), 52–65.

Davidson-Hunt, I. and Berkes, F. (2003). Learning as you journey: Anishinaabe perception of social-ecological environments and adaptive learning. *Conservation Ecology*, 8(1), 5.

Derungs, C. and Purves, R.S. (2014). From text to landscape: Locating, identifying and mapping the use of landscape features in a Swiss Alpine corpus. *International Journal of Geographical Information Science*, 28(6), 1272–1293.

Derungs, C. and Purves, R.S. (2016). Characterising landscape variation through spatial folksonomies. *Applied Geography*, 75, 60–70.

Descola, P. (2006). *Par-delà nature et culture*. Gallimard, Paris.

Fair, S.W. (1997). Inupiat naming and community history: The *Tapqaq* and *Saniniq* coasts near Shishmaref, Alaska. *The Professional Geographer*, 49(4), 466–480.

Giraut, F. and Houssay-Holzschuch, M. (2016). Place naming as dispositif: Toward a theoretical framework. *Geopolitics*, 21(1), 1–21.

Gottschlich, D., Mölders, T., Padmanbhan, M. (2017). Introduction to the symposium on feminist perspectives on human–nature relations. *Agriculture and Human Values*, 34(4), 933–940.

Gunderson, L.H. (2001). *Panarchy: Understanding Transformations in Human and Natural Systems*. Island Press, Washington, DC.

Heikkilä, K. (2014). The forest is our inheritance: An introduction to Semai Orang Asli place-naming and belonging in the Bukit Tapah forest reserve. *Singapore Journal of Tropical Geography*, 35, 362–381.

Hunn, E.S. and Meilleur, B.A. (2010). Toward a theory of landscape ethnoecological classification. In *Landscape Ethnoecology: Concepts of Biotic and Physical Space*, Johnson, L.M. and Hunn, E.S. (eds). Berghahn Books, Oxford.

Ingold, T. (2006). Rethinking the animate, re-animating thought. *Ethnos*, 71(1), 9–20.

Ingold, T. (2011). *Being Alive: Essays on Movement, Knowledge and Description*. Taylor & Francis, London.

Johnson, L.M. and Hunn, E. (2010). *Landscape Ethnoecology: Concepts of Biotic and Physical Space*. Berghahn Books, Oxford.

Jones, R. (2016). Responding to modern flooding: Old English place-names as a repository of traditional ecological knowledge. *Journal of Ecological Anthropology*, 18(1), 9.

Jones, R. and Semple, S. (eds) (2012). *Sense of Place in Anglo-Saxon England*. Shaun Tyas, Donington.

Jorgensen, S.E., Fath, B.D., Bastianoni, S., Marques, J.C., Müller, F., Nielsen, S.N., Patten, B.C., Tiezzi, E., Ulanowicz, R.E. (2007). *A New Ecology: Systems Perspective*. Elsevier, Oxford.

Kingston, D.P. (2009). Almost every place, every rock, had a name: A consideration of place-name density on King Island, Alaska. *Alternative: An International Journal of Indigenous Peoples*, 5(1), 6–25.

Latour, B. (1999). *Politiques de la nature : comment faire entrer les sciences en démocratie ?* La Découverte, Paris.

Leopold, A. (1949). *A Sand County Almanac and Sketches Here and There*. Oxford University Press, New York.

Mark, D.M., Turk, A.G., Burenhult, N., Stea, D. (2011). *Landscape in Language: Transdisciplinary Perspectives*. John Benjamins Publishing Company, Philadelphia, PA.

Martínez, G. (1989). *Espacio y Pensamiento I. Andes meridionales*. Hisbol, La Paz.

Naveh, Z. and Lieberman, A.S. (2013). *Landscape Ecology: Theory and Application*. Springer Science & Business Media, Berlin.

Noël, A. (1995). Comment exploiter la microtoponymie ? Méthodologie appliquée à l'onomastique du pays d'Othe. *Histoire, économie et société*, 14(3), 419–426.

Ostrom, E. (1990). *Governing the Commons. The Evolution of Institutions for Collective Action*. Cambridge University Press, Cambridge.

Penko Seidl, N. (2008). Significance of toponyms, with emphasis on field names, for studying cultural landscape. *Acta geographica Slovenica*, 48(1), 2008.

Plumwood, V. (2002). *Environmental Culture: The Ecological Crisis of Reason*. Psychology Press, London.

Robbins, P. (2009). *Political Ecology*. John Wiley & Sons, Chichester.

Rose-Redwood, R., Alderman, D., Azaryahu, M. (2010). Geographies of toponymic inscription: New directions in critical place-name studies. *Progress in Human Geography*, 34(4), 453–470.

Salas Carreño, G. (2016). Places are kin: Food, cohabitation, and sociality in the Southern Peruvian Andes. *Anthropological Quarterly*, 89(3), 813–840.

Samakov, A. and Berkes, F. (2017). Spiritual commons: Sacred sites as core of community-conserved areas in Kyrgyzstan. *International Journal of the Commons*, 11(1), 422–444.

Spedding, A. (1992). Almas, anchanchus y alaridos en la noche: el paisaje vivificado de un valle yungueño. In *Etnicidad, economía y simbolismo en los Andes*, Arze, S., Barragan R., Escobari, L., Medinacelli, X. (ed.). Institut français d'études andines, HISBOL, Sociedad Boliviana de Historia, Lima.

Taller de Tradición Oral and Beaucage, P. (1996). La bonne montagne et l'eau malfaisante. Toponymie et pratiques environnementales chez les Nahuas de basse montagne (Sierra Norte de Puebla, Mexique). *Anthropologie et sociétés*, 20(3), 33–54.

Villette, J. and Purves, R.S. (2018). Exploring microtoponyms through linguistic and geographic perspectives. *21th AGILE Conference on Geographic Information Science*, Lund.

Visvanathan, S. (2009). The search for cognitive justice [Online]. Available at: http://www.india-seminar.com/2009/597/597_shiv_visvanathan.htm.

Wartmann, F.M. and Purves, S.R. (2017). What's (not) on the map: Landscape features from participatory sketch mapping differ from local categories used in language. *Land*, 6(4), 79.

Wartmann, F.M. and Purves, R.S. (2018). This is not the jungle, this is my barbecho: Semantics of ethnoecological landscape categories in the Bolivian Amazon. *Landscape Research*, 43(1), 77–94.

Wartmann, F.M., Derungs, C., Purves, R.S. (2016). Characterizing place: An empirical comparison between user-generated content and freelisting data. *International Conference on GIScience Short Paper Proceedings*, 336–339.

Zadora-Rio, E. (2001). *Archéologie et toponymie : le divorce*. Les petits cahiers d'Anatole, Tours.

Zimmerer, K.S. (2000). The reworking of conservation geographies: Nonequilibrium landscapes and nature-society hybrids. *Annals of the Association of American Geographers*, 90(2), 356–369.

Zipf, G.K. (1935). *The Psychobiology of Language*. MIT Press, Cambridge, MA.

# 4

# Naming the Conquered Territories: Colonies and Empires – Beneath and Beyond the Exonym/Endonym Opposition

**Frédéric GIRAUT**
*UNESCO Chair in Inclusive Toponymy "Naming the World", University of Geneva, Switzerland*

The naming of conquered places is an activity that is part of the conquest, anticipating it, accompanying it or confirming it: naming "open" or "virgin" lands, those of discovered islands, of settlers' frontiers; naming conquered or annexed territories, or subjugated or colonized populations. Indeed, imperial or colonial conquest produces territory, either instead of, or superimposed on, or incorporating previous territorialities, which are erased or dominated. If we take Claude Raffestin's (1986) definition of territoriality, this cannot be reduced to a raw political and functional appropriation of the territory (the exercise of hegemonic power over the exploitation of resources and populations); it involves a semiotization of space, a production of meaning that refers to a culture and to references and that participates in creating them.

> Territory is a reordering of space, whose order is to be found in the information systems available to a human being as a member of a culture. Territory can be considered as space informed by the semiosphere (semiosphere = the whole set of signs); all the

*The Politics of Place Naming*,
coordinated by Frédéric GIRAUT and Myriam HOUSSAY-HOLZSCHUCH. 

> mechanisms of translation, used in relations with the exterior, belong to the structure of the semiosphere" (Raffestin 1986).

Neotoponymy (or production of new place names) through the activity of naming and renaming is part of this production of territory through semiotization.

In the case of imperial or colonial conquest, it involves the use of specific and generic technologies. These include erasure, transposition and promotion, for the names used by power, but they can also use recognition, sanctuarization and toponymic appropriation for the places constituting the territoriality of the dominated populations, or even accommodate the controlled development of indigenous names. Some research works in linguistics document and analyze different colonial practices and their rationale (Carter 1987; Van den Avenne 2012; Weber 2019; Lefkovich 2020).

Repertoires for the conquest of new spaces are in fact numerous and unfold over the duration of empires (Burbank and Cooper 2010; Deprest et al. 2011). The long-distance projection of a society in the form of settler frontiers was enacted spectacularly during the first European colonization, that of the islands of the South Seas and the New World, but it found its origins in antiquity, in particular through the colonization initiated by certain Greek cities in the Mediterranean. The conquering and annexationist empire is quite an ancient figure, represented in particular by the Roman model, which also used colonies projected over long distance, as a modality. In contrast, the Ottoman Empire represented, before its final drift, the practice of cultural autonomy and indirect exploitation. The modalities and repertoires of imperialism are thus numerous and evolve over the very long term. Therefore, it is not a question here of periodization, but rather of models, repertoires, variants, modalities and of course contexts, to identify typical *dispositifs* in their complexity and their hybridizations.

The proposal here is to present this set of imperial toponymic practices and to consider how they form different *dispositifs* according to context, and always introduce complex, dialectical relationships between names of external origin and names of local origin, between exonymy and endonymy – terms that it would be very simplistic to view as strictly opposed.

This chapter will begin by evoking two canonical models – that of long-distance colonial projection, based on the toponymic invention of the fictitious Mysterious Island and the colonization of Quebec, and that of imperialism by annexation, based on the toponymic condescension of the Ottoman millet and the Roman builder-empire. It will then consider the hybridizations, alterations and evolutions of these models (with many variations depending on the context). This makes it possible to discuss in detail the exonym/endonym opposition and its very relative value for understanding the complex processes of territorialization resulting from imperial or colonial conquests.

## 4.1. Toponymic colonization of settler frontiers (long-distance metropolitan projections): the fictitious model of the Mysterious Island and its extensions

Jules Verne's work published in 1874, entitled *The Mysterious Island*, constitutes in its Chapter 11 a veritable small treatise on colonial toponymy. A fictitious small troop of shipwrecked explorers find themselves dealing with a mysterious desert island. The island is quickly mapped from a high point; the operation is similar to the baptism and official naming of all the geographical figures that explorers, who have become settlers, recognize or invent (Tissier 1996; Dupuy 2011; Tort-Donada 2014; Jenkins 2017).

**The Mysterious Island, Jules Verne**, Extracts (part 1, ch. 11)

"If Cyrus Harding was not mistaken in his calculation, the island had almost the extent of Malta or Zante, in the Mediterranean; but it was at the same time much more irregular, and less rich in capes, promontories, points, bays, or creeks. Its strange form caught the eye, and when Gideon Spilett, on the engineer's advice, had drawn the outline, they found that it resembled some fantastic animal, a monstrous leviathan, which lay sleeping on the surface of the Pacific.

This was in fact the exact shape of the island, which it is of consequence to know, and a tolerably correct map of it was immediately drawn by the reporter [...].

The exploration of the island was finished, its shape determined, its features made out, its extent calculated, the water and mountain systems ascertained. The disposition of the forests and plains had been marked in a general way on the reporter's plan. They had now only to descend the mountain slopes again, and explore the soil, in the triple point of view, of its mineral, vegetable, and animal resources.

I say by chance, because this is an unimportant island; there is not even a port in which ships could anchor, and it is to be feared that it is situated out of the route usually followed, that is to say, too much to the south for the ships which frequent the archipelagoes of the Pacific, and too much to the north for those which go to Australia by doubling Cape Horn. I wish to hide nothing of our position from you –"

– "And you are right, my dear Cyrus," replied the reporter, with animation. "You have to deal with men. They have confidence in you, and you can depend upon on them. Is it not so, my friends?"

– "I will obey you in everything, captain," said Herbert, seizing the engineer's hand.

– "My master, always and everywhere!" cried Neb.

– "As for me," said the sailor, "if I ever grumble at work, my name's not Jack Pencroft, and if you like, captain, we will make a little America of this island! We will build towns, we will establish railways, start telegraphs, and one fine day, when it is quite changed, quite put in order and quite civilized, we will go and offer it to the government of the Union. Only, I ask one thing".

– "What is that?" said the reporter.

– "It is, that we do not consider ourselves castaways, but colonists who have come here to settle."

[...]

"One minute, my friends," said the engineer, "it seems to me it would be a good thing to give a name to this island, as well as to, the capes, the promontories, and watercourses which we can see".

"Very good," said the reporter. "In the future, that will simplify the instructions which we shall have to give and follow."

"Indeed," said the sailor, "already it is something to be able to say where one is going, and where one has come from. At least, it looks like somewhere".

"The Chimneys, for example," said Herbert.

"Exactly!" replied Pencroft. "That name was the most convenient, and it came to me quite of myself. Shall we keep this name of the Chimneys for our first encampment, captain?"

"Yes, Pencroft, since you have so christened it."

"Good! as for the others, that will be easy," returned he sailor, who was in high spirits. "Let us give them names, as the Robinsons did, whose story Herbert has often read to me; Providence Bay, Whale Point, Cape Disappointment!"

"Or, rather, the names of Captain Harding," said Herbert, "of Mr. Spilett, of Neb! –"

"My name!" cried Neb, showing his sparkling white teeth.

"Why not?" replied Pencroft. "Port Neb, that would do very well! And Cape Gideon –"

"I should prefer borrowing names from our country," said the reporter, "which would remind us of America".

"Yes, for the principal ones," then said Cyrus Harding, "for those of the bays or seas, I admit it willingly. We might give to that vast bay on the east the name of Union Bay, for example, to that large hollow on the south, Washington Bay; to the mountain upon which we are standing, that of Mount Franklin, to that lake which is extended under our eyes, that of Lake Grant; nothing could be better, my friends. These names will recall our country, and those of the great citizens who have honored it; but for the rivers, gulfs, capes, and promontories, which we perceive from the top of this mountain, rather let us choose names which will recall their particular shape. They will impress themselves better on our memory, and at the same time will be more practical. The shape of the island is so strange that we shall not be troubled to imagine what it resembles. As to the streams which we do not know as yet, in different parts of the forest which we shall explore later, the creeks which afterwards will be discovered, we can christen them as we find them. What do you think, my friends?"

The engineer's proposal was unanimously agreed to by his companions. The island was spread out under their eyes like a map, and they had only to give names to all its angles and points. Gideon Spilett would write them down, and the geographical nomenclature of the island would be definitely adopted. First, they named the two bays and the mountain, Union Bay, Washington Bay and Mount Franklin, as the engineer had suggested.

"Now," said the reporter, "to this peninsula at the southwest of the island, I propose to give the name of Serpentine Peninsula, and that of Reptile-end to the bent tail which terminates it, for it is just like a reptile's tail".

"Adopted," said the engineer.

"Now," said Herbert, pointing to the other extremity of the island "let us call this gulf which is so singularly like a pair of open jaws, Shark Gulf".

"Capital!" cried Pencroft, "and we can complete the resemblance by naming the two parts of the jaws Mandible Cape".

"But there are two capes," observed the reporter.

"Well," replied Pencroft, "we can have North Mandible Cape and South Mandible Cape".

"They are inscribed," said Spilett.

"There is only the point at the southeastern extremity of the island to be named," said Pencroft.

"That is, the extremity of Union Bay?" asked Herbert.

"Claw Cape," cried Neb directly, who also wished to be godfather to some part of his domain.

In truth, Neb had found an excellent name, for this cape was very like the powerful claw of the fantastic animal which this singularly-shaped island represented.

Pencroft was delighted at the turn things had taken, and their imaginations soon gave to the river which furnished the settlers with drinking water and near the balloon had thrown them, the name of the Mercy, in true gratitude to Providence. To the islet upon which the castaways had first landed, the name of Safety Island; to the plateau which

> crowned the high granite precipice above the Chimneys, and from whence the gaze could embrace the whole of the vast bay, the name of Prospect Heights;
>
> Lastly, all the masses of impenetrable wood which covered the Serpentine Peninsula were named the forests of the Far West.
>
> The nomenclature of the visible and known parts of the island was thus finished, and, later, they would complete it as they made fresh discoveries.
>
> [...]
>
> Everything was finished, and the settlers had only to descend Mount Franklin to return to the Chimneys, when Pencroft cried out,:
>
> "Well! we are preciously stupid!"
>
> "Why?" asked Gideon Spilett, who had closed his notebook and risen to depart.
>
> "Why! our island! We have forgotten to christen it!"
>
> Herbert was going to propose to give it the engineer's name and all his companions would have applauded him, when Cyrus Harding said simply, "Let us give it the name of a great citizen, my friend; of him who now struggles to defend the unity of the American Republic! Let us call it Lincoln Island!"
>
> The engineer's proposal was replied to by three hurrahs.
>
> And that evening, before sleeping, the new colonists talked of their absent country [...].

Most of the resources of colonial toponymy, when it falls to European explorers, are presented here to constitute a real model:

– transposition of their places of origin, of its founding myths and its great men (Lincoln Island; Union Bay; Washington Bay; Mount Franklin; Grant Lake; Far West Forest);

– celebration of the "discoverers" or their relatives (Harding; Port Neb; Cape Gideon eventually discarded here);

– celebration of the topographical and historical highlights of the "discovery" of the territory (Safety Island; Mercy River; The Chimneys; Granite House; Wreckage Point; Port Balloon);

– presentation of exotic or extraordinary configurations of places (Serpentine Peninsula; Reptile End; Shark Gulf; North Mandible Cape & South Mandible Cape; Claw Cape).

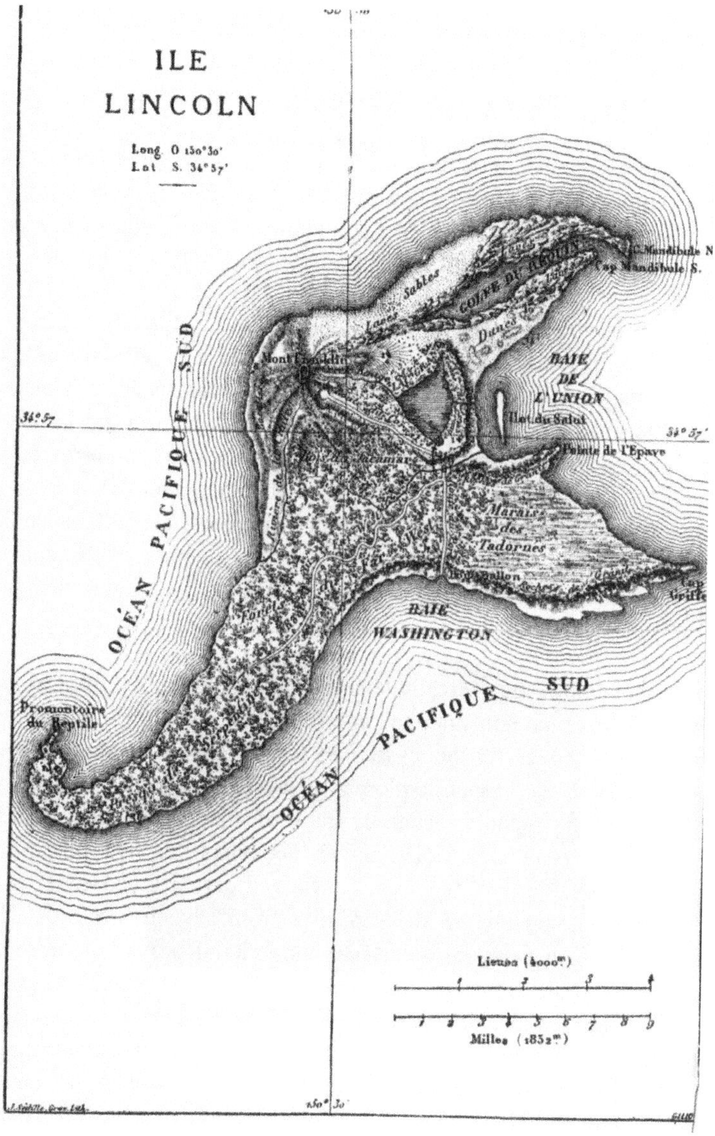


**Figure 4.1.** *Lincoln Island (illustrator J.D. Férat) in* The Mysterious Island, *Hetzel edition, 1875*[1]

1. https://fr.wikipedia.org/wiki/L%27%C3%8Ele_myst%C3%A9rieuse#/media/File:Ile_Mysterieuse_03.jpg, originally drawn by Jules-Descartes Férat (1819–1889?) – This image was originally featured in the Hetzel edition of *The Mysterious Island*, and has also been featured in more recent editions (this particular instance was scanned from a recent edition).

The model is simple, inspired by European practices in the tropics and beyond in the South Seas (Douglas 2014; Sawday 2018) during the first colonization which began in the 15th century, continuing until the 19th. It would be reproduced later for new settler frontiers. The Alpine and Pyrenean peaks, having been climbed and recognized by the pioneer mountaineers of the peaks, are thus named after their conquerors (Giraut 2019). On Mars, during the Pathfinder mission, NASA created many place names which borrow from a colonial register of transposition and give a potentially social dimension to this open space (Dittmer 2007).

In the case of the Mysterious Island, during these phases of discovery and conquest of "virgin" lands or peaks, the a priori absence of any human occupation makes the spontaneous operation of naming these new lands simple. This is already, metaphorically, an ingredient of colonial toponymy, which either largely ignores vernacular toponymy and considers it absent, if it is not imposed or remembered; or instrumentalizes it, recuperates it or relegates it (this fundamental aspect is linked to imperial practices of zoning space which will be discussed later). Because the matter becomes considerably more complicated when coming into contact with populations, it will therefore be necessary to add several essential elements to the model.

This is what another story already tells us, still relating to the tropical islands (of the South Seas), this time non-fictional, and following directly from the prototypical experience of the discovery of the mysterious island. This is the story of the naming of the Pacific Islands by French explorers in the 19th century, studied by Hélène Blais (2001), which is more generally part of the cartographic enterprise of "discoveries" (Kent et al. 2019). The author deciphers this new phase of naming which aims to produce a territory on scientific and organized bases. It involves an ordered and clear nomenclature that borrows from several registers. The initial register of the nationalist repertoire, or even more the affective repertoire (linked to the circumstances of the discovery and its effects on the discoverers) (Rose 1997), is only partially extended during these phases of description and more systematic mapping. Borrowings from selected and interpreted indigenous references are then not only admitted, but promoted as a means of objectification in the creation of a conquered geographical reality. This is part of an objective ordering of a conquered land which thereby gains in specificity and whose domination is no longer based solely on the hazards of exploration and taking possession, but on a particular place and resources in a complex geographical and political *dispositif*, an empire and its possibilities in terms of exploitation. This powerful analysis is corroborated by the work performed for other Pacific islands in the English domain for Norfolk Island (Nash 2017), as well as successively the Dutch, Chinese and Japanese at Orchid Island (Wu and Lay 2014). It is thus necessary to add a widely deployed modality to the initial model: *the partial recuperation and alteration of elements of the indigenous toponymic corpus.*

We must now add to this model of colonial motivations and practices the teachings of the practices of settler colonization, the steamroller that names its settlements, its places along the settler frontiers that it opens up. This happens in differentiated registers, depending on the cultures of the settlers. This is because their motivations differ from those of sailors and mandated explorers who initiate conquests by attesting to "discovery" and by preparing for eventual colonization by creating a coherent geographical framework. From now on, it is a question of "penetrating" a territory starting from bridgeheads, by discovering other realities and above all by creating colonial settlements, of inventing a new landscape by erasing or transforming the pre-existing order that is considered natural.

The "transposition of the places of origin" element of the initial model then massively translates into the reproduction and projection of numerous toponyms of the metropolises, with the addition of the qualifier "new". This is part of a vitalist settler impulse to invent a new world from an external origin. Apart from the so-called New Zealand, New Caledonia and New Hebrides, the naming of American cities registers a large number of toponyms of European cities preceded, or not, by the qualifier "New" (Zelinski 1967). This is the case for New Amsterdam renamed New York, as well as of the small towns called Paris to capture the prestige of a European capital, as highlighted in the film *Paris, Texas* by Wim Wenders (Reinhardt 2017). Analysis of the original denominations of plantations in Maryland in the late 17th and early 18th centuries also shows toponymic constructions modeled on the English countryside (Musselwhite 2021).

Another massive practice is *naming after biblical names* of the Old Testament, in particular for populations of Protestant origin in the United States or South Africa, and after the names of saints for populations of Catholic origin. Hagiography is thus dominant in Latin America and Quebec, a Catholic variant of the founding myths of colonial identity (Leonard 2010).

**Wajdi Mouawad with his novel *Anima*** (2012, Leméac and Actes Sud) goes beyond what words and names can say. Names are nevertheless constantly questioned in their derisory limits, as well as in their evocative power and in their (in)capacity to give meaning. Extracts which particularly deal with biblical colonial toponyms faced with the tragedies of war:

"His sister lives in Lebanon, a small town not far from Springfield, in southwestern Illinois. She keeps a *bed and breakfast* over there, The Sunrise. Her name is Ashleen. She is not married. She has kept her father's surname: Ashleen Wolf Rooney [...]".

> "Lebanon? Like the country?"
>
> "Yes, why?"
>
> He did not answer. He had lost himself in the depths of the clouds, over there, which had invaded the expanse of the sky. Something had stirred inside him (p. 210).
>
> [...]
>
> "The name on the map intrigued me. Lebanon. I came to see it."
>
> "Further south, you've got Cairo. It's worth the detour" [...].
>
> They fell silent to listen.
>
> "This region is called the Land of Egypt. The inhabitants, even today, call themselves Egyptians. The Mississippi is their Nile. In Cairo, you could admire the meeting of the waters, where the Ohio empties itself into the Mississippi. It's like that. People here, like everywhere else on American soil, turned to the Bible to baptise their land with names that would bring them happiness. In Cairo, they hoped for the coming of a Moses who would save the people from famine, disease and floods. And still further down, men possessed by a real faith in their future, remembered the great thousand-year-old city of the Pharaohs and baptised their small town Memphis. Here, Lebanon, in homage to the country where Christ performed his first miracle by multiplying bread and wine during the wedding at Cana. For people who were starving, it obviously made sense."
>
> "And that saved them?"
>
> "I would rather say that it doomed them. Calamities followed. Famines, floods, diseases and, finally, civil war. Many men died on the paths of their childhood. There's a border here, the Mason-Dixon Line. During the War of Secession, it separated the northern states from the southern ones. We were at the crossroads. Illinois was unionist and Missouri, which is ten kilometres away, was a slave state. The Union army set up a fortification in Cairo. In Lebanon too, the civil war took its toll" [...].
>
> "Many Americans have died in rivers named after them, because they were named by their ancestors. They died amid their flocks and in orchards which they should have ploughed during their lifetime. As for the French, they died far from home, helpless, lost in the midst of the Great American Civil War" (pp. 291–292).

In addition to the characteristic hagiography of toponymy in Quebec (Adam 2008), the famous novel by Louis Hémon, *Maria Chapdelaine*, expresses the affective and cultural burden conveyed by the toponymic landscape created by the settlers, in an existing context of linguistic colonial competition with the neighboring Anglophones. Colonial microtoponymy thus borrows in French from the register of the description of sites, the environment and resources. The vain attempt to eradicate all indigenous toponymy seems to stem more from the hagiographic enterprise of the Catholic Church than from the motivations of the settlers who remained in contact with certain populations and could, despite everything, admit their toponymic presence.

**Louis Hémon, *Maria Chapdelaine*,** 1921

When choosing between her two suitors who represent an exodus to the big American city or persisting in the hard life of the settlers of the Lac Saint Jean region, Maria Chapdelaine hears a voice. This reminds her in particular of the value of a linguistic landscape which is that of her people, created by the settlers and the priests who came from France, opposed to the country of English names which also end up "seeming natural" but define a somewhere-else.

**Chapter XV**

"And yet they must bc bcautiful," thought she, still dreaming of those vast American cities. As though in answer, a second voice was raised. Over there was it not a stranger land where people of an alien race spoke of unfamiliar things in another tongue, sang other songs? Here…

The very names of this her country, those she listened to every day, those heard but once, came crowding to memory: a thousand names piously bestowed by peasant from France on lakes, on rivers, on the settlements of the new country they were discovering and people as they went – lac à l'Eau-Claire – la Famine – Saint-Coeur – de Marie – Trois-Pistoles – Sainte-Rose-du-Degel – Pointe-aux-Outardes – Saint-André-de-l'Epouvante… An uncle of Eutrope Gagnon's lived at Saint-Andre-de-l'Epouvante; Racicot of Honfleur spoke often of his son who was a stoker on a Gulf coaster, and every time names were added to the old; names of fishing villages and little harbors on the St. Lawrence, scattered here and there along those shores between which the ships of the old days had boldly sailed toward an unknown land – Pointe-Mille-Vaches – les Escoumins – Notre-Date-du-Portage – les Grandes Bergeronnes – Gaspe.

How sweet to hear these names where one was talking of distant acquaintance and kinsfolk, or telling of far journeys! How dear and neighbourly was the sound of them, with a heart-warming friendly ring that made one feel as he spoke them: – "Throughout all this land we are at home… at home…"

Westward, beyond the border of the Province; southward, across the line were everywhere none but English names. In time one might learn to speak them, even might

> they at last come familiarly to the ear; but where should one find again the happy music of the French names?
>
> **Chapter XIV**
>
> "From their house to Honfleur, eight miles; from Honfleur to La Pipe, six. There her father would speak with the curé, and then pursue his way to Mistook. She corrected herself, and for the ancient Indian name that the people of the country use, gave it the official one bestowed in baptism by the church – St. Coeur de Marie. From La Pipe to St. Coeur de Marie, eight miles – eight and six and then eight. Growing confused, she said to herself – 'Anyway it is far, and the roads will be heavy.'"

With the evocation of these settler practices, their meaning and their appropriation, the model of the toponymic production of colonial projections can therefore be finalized. In addition to the urban and hagiographic variants of the transposition of places and myths of origin, there is a tendency to relegate indigenous place names by erasing or confining them to socially devalued indigenous and vernacular uses.

This parallels or goes beyond the recuperation-alteration of certain native corpora already revealed with the enterprise of mapping colonizable island domains. We might also note a variant of toponymic exoticization with derogatory references to the other, to the indigenous, to the colonized. Thus, Mark Monmonier (2006) gives the title *From Squaw Tit to Whorehouse Meadow: How Map Names Claim and Inflame* to his essay on the conflictual dimension of toponymy. These practices of colonial toponymy can be translated by the following element of the model: the inscription in the colonial toponymy of a derogatory image of the other, of the native, of the colonized.

Here, we find ourselves in the small factory of the colonial toponomascape, which is the fruit of the settlers' activity of colonization and territorialization. It is not specific to the long-distance colonial projection that accompanied the first European colonization, which was inaugurated by the European "great discoveries", their islands and stopovers on the routes to India, then subsequently the establishment of the triangular trade with its European outposts: marketplaces, plantation colonies and settler frontiers towards a Wild West. Indeed, here we find ourselves in practices which, in a different form, may merge with the toponymic colonization practiced within the framework of the continental empires which conquer to annex or exploit peripheries and to ensure their "greatness".

## 4.2. Toponymic imperialism: the model of Roman super(im)position and Ottoman condescension

The continental expansionist empire extends by the progressive conquest of its peripheries, and not by long-distance projection. It conquers the territories of the other, the neighbor, dominated and defeated, but the groups thus brought into the empire have a place; certainly, this is a subordinate and exploited one, but one in which their culture can be recognized or even protected and its cultural landscape possibly preserved, while integrating the influences and achievements of the imperial power, which are then marked in the language and the semiology of the center. Here, we enter into a wide range of repertoires or political technologies for the management of dominated diversity.

The "Romanization" of the Roman Empire took place in a progressive and differentiated way (Chambon 2002; Leroux 2004). Urban planning and spatial planning were at the heart of the enterprise, driven by an integrative project. Thus were created new centers, capitals and main towns, given equipment and infrastructures based on the technologies that the Empire was able to disseminate and transpose. These new centers were not only intended to impose a military and political order on conquered lands, as with the Alexandrias created during the Macedonian conquest, an attempt at superficial imperial unification of an empire carved out at a galloping pace. Much more than a simple technique of the military control of vast foreign and culturally autonomous entities, in the case of Rome, the project was integrative: it was therefore also administrative, economic and cultural, operating through language and religion. This consubstantial dimension to the Roman Empire was conveyed among other things by a Roman toponymy. This toponymy was imposed within these new cities and on all the related infrastructures, and spread more or less to their periphery through means such as centuriations or Roman settler colonies. A toponymic layer is therefore superimposed through the imposition of a network of centers, infrastructures and annexes, and through the administrative territorialization of the spaces of the empire. It is *the planning and naming of a network of centers and infrastructures according to the figures (political, military and religious) of imperial power*.

The period of the reign of Hadrian, a period mostly of pacification and planning between periods of trouble, offers a number of examples of the practices of a toponymic Romanization through superposition and the work of building. These include the famous walls and gate, both named after the emperor who designed the *limes* as well as the renovation of Athens; as well as the names of towns, temples and administrative divisions inspired by his young lover, his relatives or even famous Roman cultural references. This is also the case for the new city, Ælia Capitolina (in reference to the family deity of Emperor Hadrian, Ælius, and Jupiter

Capitolinus, supreme god of Rome), designed to replace Jerusalem and which triggered a resistance that was terribly repressed. Beyond the cities, the long-term mechanisms of acculturation, very different from one region of the empire to another, ensured an always partial toponymic Romanization of certain countrysides.

***Memoirs of Hadrian*, a book by Marguerite Yourcenar** (1951), recognized for its rigor and its relevance in its account and analysis of Hadrian's achievements, is very explicit about the toponymic practices and motivations of the emperor. Here are some excerpts:

"Antinoopolis was to be born; that would already be a check to death to impose upon such a sinister land a city wholly Greek, a bastion which would hold off the nomads of Erythrea, a new market on the route to India" (p. 207).

"I undertook to plan the administrative divisions of Antinoopolis, its demes, its streets, its city blocks, on the model of the world of the gods, and at the same time to include a reflection of my own life. All the deities were to be represented, Hestia, and Bacchus, divinities of the hearth and of the orgy, the gods of the heavens and those of the underworld. I placed my imperial ancestors there, too, Trajan and Nerva, now an integral part of that system of symbols. Plotina figured; the good Matidia was there, in the likeness of Demeter; my wife herself, with whom at the time my relations were cordial enough, made one of that procession of divinities. Some months later I bestowed the name of my sister Paulina upon a district of Antinoopolis; I had finally broken off with her as the wife of Servianus, but she had now died and thus had regained her unique position of sister in that city of memories. The site of sorrow was becoming the ideal center for reunions and recollections, the Elysian Fields of a life, the place where contradictions are resolved and where everything, within its rank, is equally sacred" ("Disciplina augusta", p. 227).

"We made a stop at Jerusalem. There I took occasion to study the plan for a new capital which I proposed to construct on the site of the Jewish city ruined by Titus. Good administration in Judaea and greater commerce with the Orient showed the need for developing a great metropolis at this intersection of routes. I had in mind the usual Roman capital: Aelia Capitolina would have its temples, its markets, its public baths, ad its sanctuary of the Roman Venus" ("Sæculum aureum", p. 191).

"Plotinopolis grew from the need to establish new market towns in Thrace [...]. Hadrianotherae is designed to serve as a trading town for the forest dwellers of Asia Minor [...]. Hadrianopolis in Epirus reopened an urban center in the heart of an impoverished province: it owes its start to a visit which I made to the oracle of Dodana. Hadrianopolis in Thrace, an agricultural and military outpost strategically placed on the edge of barbarian lands, is populated by veterans of the Sarmatian wars: I know at first hand the strength and the weakness of each one of those men, their names, the number of their years of service, and of their wounds Antinoopolis, dearest of all, born on the site of sorrow, is confined to a narrow band of arid soil between the river and the cliffs. I was only the more desirous, therefore, to enrich it with other resources, trade with India, river traffic, and the earned graces of a Greek metropolis" (*Memoirs of Hadrian*, p. 137).

"My estates in Africa, inherited from my mother-in-law Matidia, must be turned into models of agricultural development; the peasants of Borysthenes, the village established in memory of a good horse, are entitled to aid after a severe winter; on the contrary, subsidies should not be granted to the rich cultivators of the Nile Valley, who are ever ready to take advantage of the emperor's solicitude" (Patientia p. 291).

Thus, the Roman Empire, a builder-empire, can be seen as an integrative imperial enterprise, working intensively on the conquered territories. In this it differs from a more extensive enterprise, such as that of the Ottoman Empire, a delegating empire, and of its millet system, which freezes community affiliations while recognizing them. Each group and each territory was assigned to a culture defined by religion and potentially to a language (Bruneau 2015). In a periphery of the empire like Algeria, the imperial layer of Turkish toponymic influence is reduced to a few bridgehead cities and a possible influence on the so-called Maghzen lands (of the Empire; in other words, in the possession of local dignitaries who have pledged allegiance and can serve as relays, especially for conscription). On the other hand, the territory known as the Bled Siba (rebel, not truly conquered) and the collective village and tribal lands known as Azels, and more generally the mobile space of the tribes, are not intended to undergo a toponymic influence. So, the French found themselves in the 19th century faced with a territorial and toponymic situation that was unreadable for them (Blais 2014). Elsewhere, in the non-Muslim parts of the Empire, such as present-day Bulgaria, a certain progressive Islamization was accompanied by a partial Turkification of the toponymy of the countryside (Kiel 1992). In general, Henri Georgelin (2008) specifies that Turkish as lingua franca of the empire could be appropriated everywhere in the empire, including at the toponymic level, without it being an imperial imposition, but rather a convenience claimed for the purpose of existing within a larger whole – in other words, opting for the vehicular rather than reproducing the vernacular.

Two empires, two approaches therefore: one integrating, the other delegating, but nonlinear and non-continuous practices, also featuring convergences. Two new elements of models thus appear with respect to the dominated and dominant cultures of the empire, which complete the panoply and the extent of the conquering toponymic techniques: *the toponymic sanctuarization of conquered cultures* (particularly developed by the Ottoman Empire before its final nationalist drift), and *the promotion, appropriation and toponymic extension of a dominant imperial linguistic and religious culture* (Roman and Ottoman Empires, depending on the period and the region). Both apply to modern European settlements in their territorial phase, whether in Latin America with the relegation of indigenous cultures outside the scope of official territoriality; in North America, Australia and South Africa, with the practice of towns and domains with a very large colonial majority, on the one hand, and reduced "native reservations", on the other; or even in Algeria and in certain protectorates, with a majority indigenous domain outside of the towns and regions of agricultural colonization.

Thus, four key moments and modes of conquest – the "discovery" of tropical islands, the European creation of the New World, the Roman Empire as a builder, the Ottoman Empire's delegation – experimented with a wide variety of toponymic practices linked to the logics of imperial conquests, and provided a grammar that was widely reused through combination of the elements. The exemplary cases analyzed so far therefore make it possible to formalize a kind of general model in the form of a set of standard practices vis-à-vis the places and territories conquered and vis-à-vis the territoriality of the colonized.

**Typical elements of imperial and colonial toponymic practices**

**a) Naming "discoveries", marking conquests:**

Transposition of the places of origin, of its founding myths and of its great men

Celebration of the important topographical and historical sites of taking possession

Presentation of exotic or extraordinary configurations of places

Celebration of "discoverers" or "settlers" and their loved ones

Planning and naming of a network of centers and infrastructures after figures (political, military and religious) of imperial power

**b) Renaming the colonized and their territories**

Partial recuperation and alteration of elements of the indigenous toponymic corpus

Relegation of indigenous place names through deletion or cantonment

Toponymic sanctuarization of conquered cultures

Inclusion in the colonial toponymy of a derogatory image of the other, of the native, of the colonized

Promotion, appropriation and toponymic extension of a dominant imperial linguistic and religious culture

## 4.3. Who's in, who's out? Colonial hybridizations and relativity of the concepts of exonym and endonym

The combination of these standard practices is ultimately the norm in enterprises of conquest: it is a question of the unfolding of evolving cultural and territorial projects and adapting to power relations and complex situations. Indeed, the imperial and colonial toponymic factory is not only a matter of imposition, but is also determined by the culture and knowledge of the conquered societies and by their resistance, reception and appropriation of external toponymic contributions, even imposed ones. Indigenous populations and their toponymies are far from being passive in the process of colonial neotoponymy, for several reasons.

First of all, part of the toponymic colonial corpus comes from the "indigenous" corpus, through transcription or translation. This is because indigenous topographical knowledge is often essential for the settler, like the "indigenous" scout who guides the settler troop. Its knowledge is therefore recuperated, and imposes itself in a certain way in the new cartography and in the toponomascape in forms that may vary from the re-applied use of the toponym with a more or less faithful transcription, or a translation, or even by the creation of a tautological toponym, composed of two nouns with the same meaning in the colonial and colonized languages (Tent and Blair 2019). In his classic work, *Names on the Land*, George Rippey Stewart (1945) identifies these different types in the United States, as they were later identified and analyzed by Wilbur Zelinski (1967) and Christian Montes (2008), respectively, for the names of cities and of state capitals. In South Africa today, advocates of the Afrikaans language can argue[2] that "their" toponyms are in some way indigenous compared to those derived from English, insofar as they would much more frequently be translations of original toponyms, mainly San and Khoisan on the territory of the former Cape Colony, as well as in Bantu languages in the center (former Orange Free State) and in the north of the country (former Transvaal Republic).

But the toponymy considered as indigenous by the colonial administration may also be systematically translated into the vehicular language considered as indigenous. Various practices of arbitration for the benefit of a regional language that is more highly valued for political reasons in situations of linguistic plurality may result in a simplification and a homogenization of the official toponomascape, a true colonial creation, even though arising from an engineering based upon the cultural substrate of the colonized while reinterpreting it. Thus, the cartography and

2. Arguments heard in 2019 during a presentation and discussion at the International Symposium on Place Names, held in Clarens (South Africa), whose theme was "Recognition, Regulation, Revitalization: Place names and indigenous language."

then the administration of Algerian territory by the French initially experienced a phase of perplexity, with the perceived complexity of territories with variable geometry organized in communal areas (*azels*) and common grazing areas, and relating to historical and cultural allegiances. Soon, the temptation of simplification and categorization in a framework of modern territorialization is required. This includes a certain indigenizing unification based on Arabic as the indigenous dominant language, and some territorial normalization based on the fixity of administrative entities at the expense of the fluid space of the tribes, while the colonial domain of direct exploitation by the settlers extends.

Such practices of pseudo-recognition at the price of simplification, and parallel relegation to reserved spaces, may find their climax in the ideology and practice of apartheid or separate development in South Africa. Far from being erased, the indigenous toponymy was used there to the benefit of the policy of homelands or bantustans. In the context of the continent's decolonization, the invention of these bantustans was intended to concede pseudo-independence to the majority African populations on the basis of the transformation of reserves into pseudo-states, thereby ridding themselves of the "problems" of the civil rights of the majority while maintaining control over all resources and access to labor. The numerous parcels of land that constituted the indigenous reserves were grouped together to form "national" combinations dedicated to the different ethnic groups referenced, or even created by the regime for this program and intended to become "independent nations". Indigenous toponymy was evoked for the naming of the entities thus produced, many becoming "the land of...", modeled on the neighboring British protectorates, which served as a reference for this policy (Basutoland became Lesotho; Swaziland recently became Eswatini; Bechuanaland became Botswana; Nyasaland became Malawi). The toponymic simplification then intervened in two ways. On the one hand, the territories thus reconstituted did not correspond to the former territories of pre-colonial political entities to which they might refer (in particular Zululand or KwaZulu once reindigenized), basing their legitimacy on this pseudo-continuation. On the other hand, there was a forced regrouping by proximity of reserved parcels of land in order to constitute territories considered to be ethnically homogeneous (Christopher 1994, 2001), as well as operations of simplification during which a single language was assigned to or recognized for effectively cosmopolitan or multilingual populations and places. The townships, or reserved and segregated neighborhoods, of South African cities would then sometimes be Africanized toponymically. This was not the rule; on the contrary, generic European-language denominations (Soweto is originally the acronym for South West Townships) and dehumanizing numeration were used to differentiate roads and sectors (Pirie 1984; Guyot and Seethal 2007; Giraut et al. 2008). With the policy of Grand Apartheid, it was a question of potentially placing certain townships within the perimeter of a nearby bantustan and thus of assigning to their inhabitants the nationality of this homeland without carrying out massive relocations (as was the case in a number of situations). Thus, in Durban, the large townships of KwaMashu and Umlazi gave Zulu toponymic visibility over a vast portion of the agglomeration, while being only peripheral localities in an ocean of toponyms of European origin. Around Bloemfontein, populations considered to be Sotho were grouped together at the end of the 1970s in a vast township called Botshabelo and intended to join the bantustan of Qwaqwa and possibly be integrated into neighboring Lesotho on the basis of ethnic regrouping (Tomlinson and Krige 1997). Thus, this vast new locality of Botshabelo, isolated not far from the enclave of Thaba-Nchu (part of the bantustan of Bophuthatswana), was well separated from it, including toponymically, given its different ethno-national "destiny".

The use of indigenous or indigenous-inspired toponymy by the colonial power is therefore often made after simplification, transcription or transliteration to clearly mark, or even promote, the territories to be relegated in a system of domination and spatial segregation. But another aspect of the presence of colonized toponymy in usage, including in settler territories, is that of counter-toponymy or the use of non-indigenous vernacular toponyms. In the new post-apartheid South Africa this time, the recent practice of renaming agglomerations with names of African origin has drawn heavily on the repertoire of this counter-toponymy with the names given by the populations of African origin in the outskirts of colonial cities. Thus, Tshwane (name of a river and reference to black cows in Setswana) for Pretoria (capital named after the Afrikaner figure of Jacob Pretorius), Thekwini (reference to the figure of the bay or the power of the bull in Zulu) for Durban (named after Benjamin d'Urban, British military administrator) (Koopman 2007), while the agglomeration of Bloemfontein (fountain of flowers in Afrikaans, presented today as the translation of an indigenous toponym) has taken by metonymy and reversal of polarities the name of Mangaung (reference to the cheetah in Sesotho), after the name of the principal township of the agglomeration. The use of a counter-toponymy or unofficial double toponymy may of course be part of a process of resistance to colonization or of dual territorial claims in an asymmetrical context, as may be the case in Israel and the occupied territories in Palestine (Cohen and Kliot 1992; Azaryahu and Golan 2001; Masala 2015).

Finally, unofficial vernacular toponymy in a colonial context can be used and appropriated by the settlers themselves. We have seen an exemplary case in an excerpt from *Maria Chapdelaine*, when the heroine from the shores of Lac Saint Jean in Quebec spontaneously refers to Mistook, "the old Indian name that the locals still use," before recovering to give "to the village its official name, the one with which it had been baptised by the priests: Saint-Coeur-de-Marie". The practice of the settlers, which borrowed from indigenous toponymy, was thus thwarted or overtaken by the religious colonial administration, which imposed its own references a posteriori. Such a process leaves a certain amount of performative existence to indigenous toponymy in the hybridization of colonizing and colonized cultures.

Conversely, encounter and hybridization may also come from an indigenous toponymy which claims the space of colonial origin. Colonial urbanization engendered processes of polarization and migrations, initially limited during the colonial era, then often massive, of the (ex)colonized towards exogenous colonial towns. This urbanization, once it results in the establishment of indigenous neighborhoods, generates its own toponymy, often informal and therefore vernacular, which borrows from indigenous languages and their cultural repertoire in a great creativity linked to colonial urban life and its landmarks. The chapters by Liora Bigon and Michel Ben Arrous and the one by Melissa Wanjiru in this book give a particularly good account of this phenomenon in contemporary African urban toponymy. We thus have a hybrid toponomascape with several toponymic and odonymic layers which overlap and succeed one another (Bigon 2016). In official toponymy, there is a strong contrast between the center, where the odonymy very often retains traces of the colonial heritage in the same way as its architecture (Yeoh 1992; D'Almeida-Topor 1996; Goerg 2006; Bigon 2008; Snodia et al. 2010; Giraut and Antheaume 2012) and peripheries with increasingly marked indigenous references, particularly in the names of neighborhoods. But the vernacular toponymy in use everywhere, often unofficial, often mixes references of contrasting origin, abandoning the linear markers of the roads to attach itself to landmarks which are notable from a practical point of view.

Considering these dynamics of toponymic hybridization of different natures which inevitably mark the conquered spaces, it is the relevance of the exonym/endonym distinction in these contexts which is in question. The distinction, at the international level, is well codified and serves to standardize uses from a conventional perspective. It thus opposes names designating the same geographical object, the exonym doing so in an external language foreign to the context of the toponym, and the endonym doing so in the official context and language of the toponym (Ormeling 1993; Bartos-Elekes 2008; Woodman 2012). Drago Kladnik (2009) and Peter Jordan (2015) address the double notion in multilingual contexts where the simple opposition is equivalent to giving a patent of indigeneity to a language in a multiple toponymy and settlement. Official definitions[3] given by the

3. The glossary of terms for the standardization of geographical names, published in 2002 in four languages by the United Nations, gives the following definitions:

Genung (United Nations Group of Experts for the Standardization of Geographical Names) to the notions of exonyms and endonyms make it possible to take these difficulties into account. But in the case of spaces marked by colonization, the distinction between exonyms/endonyms made by linguists makes a distinction between names according to their colonial or indigenous origin (Stolz and Warnke 2016). However, this distinction, in many cases, is debatable as we have seen, with the many processes of hybridization. We can try to make the distinction on a semantic or etymological or even phonological basis, which would allow us to associate transcribed or even translated toponyms in the category of endonyms. But this does not take into account the encounter or the hybridity of which these names, their writing and their pronunciation are the fruit.

Nor does it account for the complex linguistic landscapes in which several languages coexist and/or are in competition, including one or more languages of colonial origin, and for which the simple opposition of exonyms/endonyms is much too simplistic, the endonym of some being the exonym of others. Algeria again constitutes an exemplary case here. Slimane Zeghidour (2015) notes in his commentary on the book by Hélène Blais (2014): "The French transplant onto Algerian soil remains so significant that Arabic place names made French by the occupier and transcribed into Latin letters ignoring the gutturals have been maintained as they are by the Algerian administration, despite a tireless effort of frenzied Arabization." We thus see how the acceptance of supposedly indigenous toponymies by the colonist operates an alteration by transcription, which is then the foundation of an inherited hybrid toponomascape which does not fit into the dual categories of exonyms and endonyms.

At the antipodes, in the Norfolk archipelago to the east of Australia, a borderline case of chronological reversal between corpora in European, Creole and regional languages is documented by Joshua Nash (2017). The names of topographic elements in Norfolk (the creole language of the archipelago, derived from English and Tahitian) of Phillip Island (one of the two uninhabited islands of the Norfolk archipelago, making three with the eponymous main island) were given from 1856 onwards. This is the date of the settlement of the main island by a group entirely

Exonym: name used in a specific language for a geographical feature situated outside the area where that language has official status, and differing in its form from the name used in the official language or languages of the area where the geographical feature is situated.

Minority name: toponym in a minority language.

Endonym: name of a geographical feature in one of the languages occurring in that area where the feature is situated.

Standardized endonym: endonym sanctioned by a names authority (glossary of terms for the standardization of geographical names).

relocated from Pitcairn Island (Central Pacific) and considered to be descendants of the crew of the *Bounty*. The author who has studied and historicized the toponymy of the archipelago believes that the English microtoponyms of Phillip Island, considered less authentic, date from the first explorations and the settlements of convicts. They would therefore be older and, according to the author, closer to the topographical and environmental realities of the island. As for the Melanesian names in the Mota language, they would in fact constitute the very last toponymic layer, promoted by the Melanesian Mission and its school installed on the archipelago for five decades straddling the 19th and 20th centuries. In New Caledonia, the origin of settlement and the origin of the toponymy are not directly linked. This is explained by the recuperation of Melanesian names to name certain settler settlements and by the toponymic work carried out by the missions in contact with the populations (Chatelier 2007). The New Caledonian toponymic paradox therefore sees Christian European names used to name separatist areas, and Kanak names naming Caldoche areas, with referents that have become territorial and community-based independent of their origin.

But it is perhaps the question of the status of the toponyms designating or having designated the current Greenland, marked by colonial toponymy (Schuster 2019), which is most telling about the confusion around the notions of endonyms and exonyms in a colonial context. The name, however, exogenous and old, of Greenland could thus appear at the end of the 20th century as an inclusive alternative to the name Kalaallit Nunaat ("Kalaallit country") which imposed itself from the 19th century as the indigenous name of the country before the Norwegian name of Greenland (Kleivan 1977). A double appellation had already been in operation during the phases of Viking and Norwegian colonization prior to Danish colonization. Greenland is a nomination attributed (from sagas written later, the thesis is therefore debated) to the Viking Erik the Red, in a possible attempt at toponymic promotion of these lands dedicated to colonization among the peoples of Iceland. During the subsequent phases of Norwegian colonization, the name Nunarput functioned in parallel to designate "the country of the natives", so called in relation to foreigners and therefore for external purposes. The shift to the expression Kalaallit Nunaat in the first part of the 20th century, during the Danish colonization, took place in a context where colonization involved the South and the West but much less the East and the North and their populations. The term Kalaallit for Greenlanders is thus an ethnic reference to South Greenlanders alone, who would have imposed this reference in the colonial context through the school and the catechism introduced by the Europeans, but within which they exercised a hegemonic cultural influence as the Indigenous people. Inge Kleivan speaks of an example of cultural imperialism exercised among the Greenlanders themselves (1977, p. 199). The expression "Nunaat Inuit" referring to all the Eskimo peoples would have been, according to the author, more inclusive and therefore acceptable by all the Indigenous people of Greenland. But this would be an extensive designation likely to apply to all Inuit lands far beyond Greenland or Kalaallit Nunaat. Since then, the autonomous territories of Canada and Quebec dedicated to the Inuit peoples have been called Nunavut ("our land" in Inuktitut) and Nunavik ("the big land" in Nunavimmiutitut, a dialect of Inuktitut).

In conclusion, we might say that the numerous cases, which testify to an almost systematic hybridization of the toponymy of conquest, should in no way lessen the diagnosis of colonial dispossession. This, based on a set of recurring practices that may operate in combination or successively (Giraut and Houssay-Holzschuch 2016), is similar to the imposition of a cultural hegemony that destroys culture, rights and knowledge (Berg and Vuolteenaho 2009). On the other hand, the study of hybridizations from a relational, pragmatic and procedural perspective makes it possible to grasp, beyond a primary opposition, the historical power of the colonial fact, a major transformer of culture for all the societies involved, dominated and dominant, victims and authors, and for those later on who issue from and are heirs to these stories. It also serves to underscore the urgency and importance of policies of restitution and recognition, particularly in situations of settler colonization, while historicizing the colonial cultural fabric more broadly.

## 4.4. References

Adam, F. (2008). L'autorité et l'autre. Parcours toponymiques et méandres linguistiques au Québec. *L'Espace politique*, 2, 31–39.

Azaryahu, M. and Golan, A. (2001). (Re)naming the landscape: The formation of the Hebrew map of Israel 1949–1960. *Journal of Historical Geography*, 27(2), 178–195.

Bartos-Elekes, Z. (2008). The discussion on terminology of the terms exonym and endonym. *Review of Historical Geography and Toponomastics*, III(5–6), 53–62 [Online]. Available at: https://geografie.uvt.ro/wp-content/uploads/2015/06/12_Bartos.pdf.

Berg, L.D. and Vuolteenaho, J. (eds) (2009). *Critical Toponymies: The Contested Politics of Place Naming*. Ashgate, Aldershot.

Bigon, L. (2008). Names, norms and forms: French and indigenous toponyms in early colonial Dakar, Senegal. *Planning Perspectives*, 23(4), 479–501.

Bigon, L. (ed.) (2016). *Place Names in Africa: Colonial Urban Legacies, Entangled Histories*. Springer, Cham.

Blais, H. (2001). Comment trouver le "meilleur nom géographique" ? Les voyageurs français et la question de la dénomination des îles océaniennes au XIXe siècle. *L'Espace géographique*, 4, 348–357.

Blais, H. (2014). *Mirages de la carte. L'invention de l'Algérie coloniale.* Fayard, Paris.

Bruneau, M. (2015). *De l'Asie Mineure à la Turquie. Minorités, homogénéisation ethno-nationale, diasporas*, CNRS. Éditions, Paris.

Burbank, J. and Cooper, F. (2010). *Empires in World History: Power and the Politics of Difference.* Princeton University Press, Princeton.

Carter, P. (1987). *The Road to Botany Bay.* University of Minnesota Press, Mineapolis.

Chambon, J.-P. (2002). Linguistique historique et archéologie : aspects toponymiques de la romanisation de la Gaule à la lumière de travaux récents concernant la Grande Limagne. *Bulletin de la Société de Linguistique de Paris*, 97(1), 95–121.

Chatelier, J. (2007). La révision toponymique (et cartographique) en Nouvelle-Calédonie (1983–1993). *Journal de la Société des Océanistes*, 125, 295–310.

Christopher, A.-J. (1994), *The Atlas of Apartheid.* Routledge, London.[4]

Christopher, A.-J. (2001). *Atlas of Changing South Africa.* Routledge, London.

Cohen, S.B. and Kliot, N. (1992). Place-names in Israel's ideological struggle over the administered territories. *Annals of the Association of American Geographers*, 82(4), 653–680.

D'Almeida-Topor, H. (1996). Le modèle toponymique colonial dans les capitales de l'Ouest africain francophone. In *La Ville européenne outre mers : un modèle conquérant ?*, Coquery-Vidrovitch, C., Goerg, O. (eds). L'Harmattan, Paris.

Deprest, F., Blais, H., Singaravélou, P. (2011). *Territoires impériaux. Une histoire spatiale du fait colonial.* Publications de la Sorbonne, Paris.

Dittmer, J.N. (2007). Colonialism and place creation in Mars Pathfinder media coverage. *Geographical Review*, 97(1), 112–130.

Douglas, B. (2014). Naming places: Voyagers, toponyms, and local presence in the fifth part of the world, 1500–1700. *Journal of Historical Geography*, 45(0), 12–24.

Dupuy, L. (2011). Une métaphore de la démarche géographique et de l'histoire du XIXe siècle: *L'Île Mystérieuse* de Jules Verne (1874–1875). *Cybergeo: European Journal of Geography.* DOI: http://dx.doi.org/10.4000/cybergeo.24646.

Georgelin, H. (2008). Thrace orientale et Anatolie : territoires à nommer et à saisir, à la fin du XIXe et au XXe siècle. In *Nommer et classer dans les Balkans*, Sintes, P. and de Rapper, G. (eds). École française d'Athènes, Athens.

Giraut, F. (2016). Changements de noms de pays africains en contexte postcolonial [Online]. Available at: https://neotopo.hypotheses.org/666.

Giraut, F. (2019). L'affaire des pics aragonais : décolonisation des sommets ou ascension de la néo-autochtonie? [Online]. Available at: https://neotopo.hypotheses.org/900.

4. Christopher (2001) is a new edition of Christopher (1994), with a title change.

Giraut, F. and Antheaume, B. (2012). Toponymy of power, power of toponymy? Colonial and contemporary Togolese place renaming. In *5th Trend in Toponymy Conference*, Bern [Online]. Available at: http://archive-ouverte.unige.ch/vital/access/manager/Repository/unige:22883?term=giraut.

Giraut, F. and Houssay-Holzschuch, M. (2016). Place naming as dispositif: Toward a theoretical framework. *Geopolitics*, 21(1), 1–21.

Giraut, F., Guyot, S., Houssay-Holzschuch, M. (2008). Enjeux de mots : les changements toponymiques sud-africains. *L'Espace géographique*, 37(2), 131–150.

Goerg, O. (2006). Domination coloniale, construction de "la ville" en Afrique et dénomination. *Afrique histoire*, 5(1), 15–45.

Guyot, S. and Seethal, C. (2007). Identity of place, places of identities: Change of place names in post-apartheid South Africa. *South African Geographical Journal*, 89(1), 55–63.

Jenkins, E. (2017). The naming of uninhabited islands: Jules Verne's *The Mysterious Island* and Marion Island. *South African Geographical Journal*, 99(2), 184–195.

Jordan, P. (2015). The endonym/exonym divide from a cultural-geographical point of view. In *Challenges in Synchronic Toponymy/Défis de la toponymie synchronique*, Löfström, J. and Schnabel-Le Corre, B. (eds). Narr Francke Attempto Verlag, Tübingen.

Kent, A.J., Vervust, S., Demhardt, I.J., Millea, N. (eds) (2019). *Mapping Empires: Colonial Cartographies of Land and Sea*. Springer, Cham.

Kiel, M. (1992). La diffusion de l'Islam dans les campagnes bulgares à l'époque ottomane (XV$^{e}$–XIX$^{e}$ siècles) : colonisation et conversion. *Revue des mondes musulmans et de la Méditerranée*, 66, 39–53.

Kladnik, D. (2009). Semantic demarcation of the concepts of endonym and exonym. *Acta Geographica Slovenica*, 49(2), 394–414.

Kleivan, I. (1977). Place names in Greenland: Cultural imperialism and cultural identity. *Transactions of the Finnish Anthropological Society*, 2, 197–215.

Koopman, A. (2007). The names and the naming of Durban. *Onoma*, 42, 73–88.

Léonard, C.J. (2010). D'une hagionymie abondante à une hiéronymie supplétive. Usage d'une stratégie toponymique chez les Franco-Canadiens de la Saskatchewan. *Cahiers de Géographie du Québec*, 54(151), 25–40.

Lefkovich, N. (ed.) (2020). *Advances in Comparative Colonial Toponomastics*. De Gruyter, Berlin.

Masala, N. (2015). Settler-colonialism, memoricide and indigenous toponymic memory: The appropriation of Palestinian place names by the Israeli state. *Journal of Holy Land and Palestine Studies*, 14(1), 3–57.

Monmonier, M.S. (2006). *From Squaw Tit to Whorehouse Meadow: How Maps Name, Claim, and Inflame*. University of Chicago Press, Chicago.

Montes, C. (2008). La toponymie comme révélateur de la construction identitaire d'un empire : (re)nommer les capitales étatsuniennes. *L'Espace géographique*, 37(2), 106–116.

Musselwhite, P. (2021). Naming plantations: Toponyms and the construction of the plantation system in the English Atlantic. *Journal of Social History*, 54(3), 741–774 [Online]. Available at: https://doi.org/10.1093/jsh/shaa020.

Nash, J. (2017). Linguistics, geography, and the potential of Australian island toponymies. *Australian Geographer*, 48(4), 519–537.

Nyambi, O. and Makombe, R. (2019). Beyond seeing QwaQwa, "Homelands", and "Black States": Visual onomastic constructions of Bantustans in apartheid South Africa. *African Studies Quarterly*, 18(4), 1–20.

Ormeling, F. (1993). Exonyms in cartography. UNGEGN training course in toponymy for southern Africa. Department of geography, University of Pretoria [Online]. Available at: http://lazarus.elte.hu/cet/modules/toponymy-2012/_HtmlModules/_Documents/D13/Documents/D13-01_Ormeling.pdf.

Pirie, G.H. (1984). Letters, words, worlds the naming of Soweto. *African Studies*, 43(1), 43–51.

Raffestin, C. (1986). Écogenèse territoriale et territorialité. In *Espaces, jeux et enjeux*. Auriac, F. and Brunet, R. (eds). Fayard, Paris.

Reinhardt, W. (2017). Quand l'Europe nomme le monde. In *Europa, Notre histoire*, François, É. and Serrier, T. (eds). Les Arènes, Paris.

Rose, G. (1997). Situating knowledges: Positionality, reflexivities and other tactics. *Progress in Human Geography*, 21(3), 305–320.

Sawday, J. (2018). They shall no more be remembered by their name: Cartography, anatomy, and the Renaissance eponym. *Journal of Medieval and Early Modern Studies*, 48(1), 11–40.

Schuster, S. (2019). The making of Greenland – Early European place names in Kalaallit Nunaat. *The Linguistic Heritage of Colonial Practice*, 13, 43.

Snodia, M., Muguti, T., Mutami, N. (2010). Political dialoguing through the naming process: The case of colonial Zimbabwe (1890–1980). *Journal of Pan African Studies*, 3(10), 16–31.

Stewart, G.R. (1945). *Names on the Land: A Historical Account of Place-Naming in the United States*. Random House, New York.

Stolz, T. and Warnke, I.H. (2016). When places change their names and when they do not. Selected aspects of colonial and postcolonial toponymy in former French and Spanish colonies in West Africa – The cases of Saint Louis (Senegal) and the Western Sahara. *International Journal of the Sociology of Language*, 239, 29–56.

Tent, J. and Blair, D. (2019). A clash of names: The terminological morass of a toponym class. *Names*, 67(2), 65–77.

Tissier, J.-L. (1996). L'île mystérieuse – Jules Verne – 1874 Hydrographie et orographie. L'île est-elle habitée ? Baptême des baies, caps, golfes, rivières. *Cybergeo: European Journal of Geography*. DOI: https://doi.org/10.4000/cybergeo.219.

Tomlinson, R. and Krige, S. (1997). Botshabelo: Coping with the consequences of urban apartheid. *International Journal of Urban and Regional Research*, 21(4), 691–705.

Tort-Donada, J. (2014). Some observations on the value of onomastics in the work of Jules Verne: A geographical perspective. In *Unconventional Anthroponyms: Formative Patterns and Decisive Function*, Felecan, O. and Felecan, D. (eds). Cambridge Scholars, Cambridge.

Van den Avenne, C. (ed.) (2012). *Linguistique et colonialisme. Communiquer, décrire, imposer*. Glottopol. Université de Rouen, Rouen.

Yeoh, B.S.A. (1992). Street names in colonial Singapore. *Geographical Review*, 82(3), 313–322.

Weber, B. (ed.) (2019). *The Linguistic Heritage of Colonial Practice*. Walter de Gruyter GmbH & Co KG, Berlin.

Woodman, P. (ed.) (2012). *The Great Toponymic Divide. Reflections on the Definition and Usage of Endonyms and Exonyms*. Geodesy and Cartography, Warszawa.

Wu, C.-J. and Lay, J.-G. (2014). Colonial powers and geographic naming: A case study of orchid island (Lanyu), Taiwan. In *History of Cartography*, Liebenberg, E., Collier, P., Török, Z.G. (eds). Springer, Berlin.

Zeghidour, S. (2015). Note de lecture sur Mirages de la Carte. L'invention de l'Algérie coloniale [Online]. Available at: https://www.iris-france.org/note-de-lecture/smirages-de-la-carte-linvention-de-lalgerie-coloniale/.

Zelinski, W. (1967). Classical town names in the United States: The historical geography of an American idea. *Geographical Review*, 57(4), 463–495.

# 5

# "Addressing the World": A Political Genealogy of the Street Address

**Reuben ROSE-REDWOOD[1], Anton TANTNER[2] and Sun-Bae KIM[3]**

[1] *Department of Geography, University of Victoria, Canada*

[2] *Institut für Geschichte, Universität Wien, Vienna, Austria*

[3] *Chungju Girls' High School, South Korea*

## 5.1. Introduction

On March 22, 2016, a demolition crew followed their GPS navigation system to what they thought was 7601 Cousteau Drive in a town near Dallas, Texas, and set to work demolishing a house, only to later discover that they had destroyed the wrong home a block away at 7601 Calypso Drive instead (Coleman 2016). Not only had the GPS system misdirected the demolition workers to the wrong street address, but the street sign indicating the intersection of Cousteau Drive and Calypso Drive had previously been knocked down following a tornado, further contributing to the confusion. As this case of "mistaken identity" makes abundantly clear, systems of street addressing have become integral features of everyday life in many places, often remaining largely unquestioned until something goes awry. Yet, it is precisely in such exceptional moments that the importance of the street address as a governmental technique of spatial identification becomes evident.

The practice of street addressing – that is, the assigning of house numbers and street names to specific locations – is generally assumed to have its origins in the history of postal communications to facilitate the delivery of mail. It may come as a

*The Politics of Place Naming*,
coordinated by Frédéric GIRAUT and Myriam HOUSSAY-HOLZSCHUCH. 

surprise, then, to find that the street address has a *political* genealogy emerging from its early use as a political technology of military logistics, police power, census-taking and tax administration. Although the historical emergence of house numbering dates back at least to the 15th century (Pronteau 1966), the affixing of numbers to buildings and the installation of street name signs at intersections did not become a widespread practice in Europe and its colonial spheres of influence until the 18th century (Tantner 2007, 2009, 2015; Rose-Redwood 2008, 2012a and Rose-Redwood and Tantner 2012).

The use of street addressing as a mode of spatial ordering is now a commonplace feature of daily life for many urban residents around the world. However, street addressing is by no means a universal practice globally, and it is estimated that "[t]here are still at least tens of millions of city dwellers around the world who don't have an address" (Mask 2018; also, see Mask 2020). In recent decades, there have been growing efforts to extend "Western-style" forms of street addressing in the name of emergency preparedness, international development and globalization. With the rise of geospatial technologies, street addressing systems have also come to serve as a basis for geocoding the world through the use of geographic information systems (GIS). At the same time, some high-tech companies have sought to bypass street addressing altogether by developing digital geocoding systems that do not rely on conventional street addresses at all, which has been particularly enticing in the Global South where formal street addressing systems often do not extend to informal settlements. The spatial histories of street addressing and related practices of geocoding are therefore part of an ongoing process that continues to reshape the geographical imaginaries and spatial infrastructures of the 21st century.

In this chapter, we provide an overview of the political genealogy of the street address. First, we consider the curious lack of scholarship on the spatial histories of street addressing and discuss the small but growing body of literature on the topic. Next, we trace the historical emergence of street addressing practices in different geographical contexts based upon the current state of knowledge. Finally, we conclude by considering potential avenues of future research on the role of street addressing in the making of a geocoded world.

## 5.2. Street addressing as a technology of power

Despite its consequential role in the spatial history of modernity, few scholars have seriously considered the history of street addressing as a technology of power. As historian Marco Cicchini explains, "house numbers have become so taken for granted that the history of their emergence has gone largely unexamined" (2012, p. 615). Smail (2000, p. 36) accounts for this lack of attention by observing that "[t]he oversight is understandable, for the address has none of the innate appeal of race, gender, madness, or sexuality. But there are good reasons for studying the historical construction of the address". Chief among them, Smail notes, is that "addresses are favored in the classificatory schemes of modern record-keeping bureaucracies because they make it possible to find people" since these "bureaucratic classifications create identities that are cellular and assemblable, identities that are useful for the nation-building project" (2000, p. 37). Viewed in this light, street addressing is a governmental strategy for producing spatial legibility to facilitate the governing of populations, places and territories.

As a mundane part of daily life, street addressing is often taken to be an apolitical practice of governmental administration. To be sure, it is widely recognized that the (re)naming of streets – and other public places – can be a highly contentious political affair, and there is a sizable body of scholarly literature on the cultural politics of street naming (Berg and Vuolteenaho 2009; Rose-Redwood et al. 2018). Much of this literature focuses on the role that street naming plays in the spatialization of historical memory through the production of commemorative landscapes. Yet, while numerous studies have examined the commemorative significance of street naming, there is very little scholarship on the historical geographies of house numbering and street addressing systems as *geo-locational regimes of spatial identification*. Such geo-locational regimes have come to serve as the condition of possibility for a wide range of political and economic practices – including everything from voter registration to opening a bank account – which makes it all the more necessary to critically examine the political genealogy of the street address.

Technical studies of different house numbering and street addressing systems have appeared in civil engineering and urban management periodicals for over a century (for a bibliographic reference of early works, see Goodman 1954), but it was not until the mid-20th century that other scholars began to take an interest in the history of house numbering (Morbito 1965; Pronteau 1966). Throughout the remainder of the 20th century, only a handful of studies extended this early work (Thale 1986; Garrioch 1994; Hunt 1995; Henkin 1998). Over the past two decades, however, a number of scholars have advanced our understanding of the "political life" of house numbering and street addressing as modes of spatial ordering (Smail 2000; Garrioch 2002; Joyce 2003; Curry et al. 2004; Farvacque-Vitkovic et al. 2005; Rose-Redwood 2006, 2008, 2012a, 2012b; Tantner 2007, 2009, 2015; Wittstock 2008; Njoh 2010; Cicchini 2012, 2018; Harris and Lewis 2012; Rose-Redwood and Tantner 2012; Vuolteenaho 2012; Rose-Redwood and Kadonaga 2016; Mask 2020; Sasse 2021).

These recent works have considered the history of street addressing from a variety of different disciplinary perspectives. Some scholars have sought to trace the early beginnings of house numbering in Europe (e.g. Tantner 2009), while others have examined more recent uses of street addressing as part of international development initiatives in sub-Saharan Africa (e.g. Farvacque-Vitkovic et al. 2005). Various studies have drawn upon Foucault's (2007) theorization of governmentality to consider how the practices of house numbering and street addressing relate to broader discussions of space, knowledge and power (Rose-Redwood 2006, 2012a; Cicchini 2012; Rose-Redwood and Tantner 2012; Tantner 2015). Although Foucault does not discuss the history of street addressing in any depth in his genealogy of modern governmentality, he suggests that the "problem of the series" – that is, keeping track of, and regulating, mobile and stationary objects – was a central concern that led to the proliferation of apparatuses of security. Foucault refers to this issue as a question of governing "an indefinite series of accumulating units: how many inhabitants, how many houses, and so on", and he maintains that "the management of these series… is pretty much the essential characteristic of the mechanism of security" (2007, p. 20). The numbering of houses and naming of streets are therefore regulatory strategies devised to produce a governable space that can serve as a "multivalent and transformable framework" to account for "a series of possible events" (Foucault 2007, p. 20).

As a governmental technique devised at the sovereignty–discipline–security nexus, house numbering is part of what Foucault calls the "urbanization of the territory", which "involved making the kingdom, the entire territory, into a sort of big town; arranging things so that the territory is organized like a town, on the model of a town, and as perfectly as a town" (2007, p. 336). The process of spatially arranging and organizing the territory "like a town" was anything but "perfect" and remains an incomplete political project even today. Yet, over the past half century, there have been numerous concerted efforts to extend city-style street addressing into the countryside to "enhance" the security apparatus of the state (Curry et al. 2004; Rose-Redwood 2012a).

Cicchini (2012) contends that the ideal of urban transparency associated with the historical emergence of urban house numbering during the 18th century contributed to the formation of a new "regime of visibility" (615). This aspiration not only underpinned philosophical and political discourses of the Enlightenment era but also informed the "street-level policing of the city" (Cicchini 2012, p. 615). Such a regime of visibility has served as the basis of what Scott (1998) calls "seeing like a state", although the political technologies of the "urban" are not reducible to the scopic regimes of state sovereignty alone (Magnusson 2011). The construction and maintenance of street addressing systems have involved both state and non-state actors, and these efforts to produce spaces of legibility have not gone uncontested historically or in the present (Rose-Redwood 2008).

## 5.3. Genealogies of the street address

The search for a "primal scene" of the origin of house numbering is likely a fruitless endeavor, since there does not appear to have been a single "inventor" of the street address. As Marx long ago observed, "a critical history of technology would show how little any of the inventions of the eighteenth century are the work of a single individual" (1990, p. 493). This certainly seems to have been the case with the history of street addressing. Keeping this caveat in mind, one of the earliest instances of house numbering can be found in 15th-century Paris, where 68 houses on the Pont Notre-Dame were numbered as early as 1463 (Pronteau 1966). Yet, in this case, some have suggested that "[t]he purpose of the numbering was not so much to provide the houses with an address as it was to keep a tally of the city's wealth" (Tantner 2015, p. 15; also, see Pronteau 1966; Tantner 2009). Another early example of house numbering is that of buildings in a social housing complex known as the Fuggerei in Augsburg by 1519, although, again, it is not known whether these numbers were initially conceived as street addresses in the modern sense.

It was not until the 18th century that house numbering plans proliferated across Europe and beyond (Tantner 2015). There is some indication that houses were numbered along Prescot Street in London in 1708, but it was not until 1765 that the Court of Common Council of London passed a law requiring the numbering of buildings and the installation of street name signs at intersections (Heal 1942; Joyce 2003). In 1727, houses and individual units within buildings were numbered in Prague's Jewish quarter as part of the *Judenkonskription* – that is, the registration and surveillance of the city's Jewish population. During the 1720s, houses in the suburbs of Paris were also assigned numbers to aid in conducting a census.

In other contexts, the introduction of house numbering was explicitly for military purposes. For instance, in 1737, Prussian authorities ordered the numbering of houses on "the day before the military enter small towns" in order to facilitate the billeting of soldiers (quoted in Tantner 2015, p. 17). In France, the quartering of soldiers among civilians was also the primary reason behind a royal decree in 1768 that required numbers to be affixed to all houses in cities and towns throughout the kingdom, with the exception of Paris. In 1749, a house numbering scheme for Paris was proposed by a French police lieutenant, but the plan was not adopted. Three decades later, a French publisher of city guides and directories led an effort to number buildings in Paris, which was eventually blocked by the police, yet house numbering soon became commonplace throughout the city in the years that followed (Pronteau 1966; also, see Rose-Redwood 2008).

The French Revolution and the Revolutionary Wars had a profound effect on the diffusion of house numbering across Europe. During this period, the French imposed systems of house numbering in the towns they occupied in Prussia (Germany), Switzerland and the Netherlands, extending the 1768 decree to occupied territories. After having considered the introduction of house numbering to enhance tax collection, "conscription numbers" were affixed to buildings in both urban and rural areas throughout the Austrian and Bohemian provinces of the Habsburg Monarchy during the 1770 census with the primary aim of preparing a new military recruiting system (Tantner 2015).

As the political technology of house numbering became more prevalent, there were various cases of resistance to early house numbering campaigns. For example, when houses were numbered to facilitate military registration in the town of Litomyšl, Bohemia, in 1770, over a dozen house numbers "were partly befouled with dung, partly gouged out using an iron bar" (as quoted in Tantner 2015, p. 24; also, see Rose-Redwood and Tantner 2012). Similarly, when house numbering was introduced in Geneva in the 1780s with the aim of "keeping order", the governing authorities encountered considerable resistance. In 1782, "nearly 150 numbers that had already been painted on the houses were erased, despite the deployment at night of military patrols which were given the task of preventing sabotage of the house-numbering operation" (Cicchini 2012, p. 620). Governmental authorities employed house numbering as part of a new regime of visibility and legibility, yet those who resisted such efforts sought to remain invisible and illegible to the apparatuses of governmentality.

By the end of the 18th century, the use of house numbering had spread beyond the European continent. Following the American Revolution, cities in the United States began adopting house numbering systems, with Philadelphia leading the way. In 1785, the publisher of a Philadelphia city directory took it upon himself to develop a numbering system based on the English practice of numbering houses along one side of the street and then continuing back along the opposite side of the same street. This sequential method of numbering caused confusion because high and low number were side by side, and it was replaced with an odd/even numbering system in Philadelphia when the first federal census was conducted in 1790 (Alotta 1990). Devised by a federal census-taker, this new system separated odd and even numbers on different sides of the street. In 1793, New York City adopted its first house numbering law modeled on the Philadelphia plan, and the odd/even system soon spread to other cities in the U.S. and abroad (Rose-Redwood 2012b).

City directory publishers played an important role in promoting the introduction of house numbering in U.S. cities during the later 18th and 19th centuries (Rose-

Redwood 2008). Directory publishers commonly petitioned city officials to initiate house numbering, because it simplified the process of linking individual urban residents and businesses to specific spatial locations in the city, thereby enhancing the *economization of time through the rationalization of space.* This goal of economizing time by rationalizing space served a wide variety of ends, as indicated by the publisher of the 1878 edition of *Trow's New York City Directory*:

> [The city directory] aids the officer of justice to apprehend the culprit or to obtain the witness, brings instantly the friend to the presence of the friend he needs, increases the usefulness of the post-office and the telegraph, facilitates the operations of commerce, and... [l]et any one who has never thought what a Directory is, try to fancy what a city of more than a million souls would be without one, and what a paralysis of social and commercial communication would be induced by the abrupt cessation of an enterprise like the Trow New York City Directory... The Directory being one of the great enterprises close of kin to the railroad, the telegraph, and the steamship, for the saving of time – time in a highly organized community being indeed money – it follows inevitably that its compilers have assumed a great responsibility... It must be nearly perfect, or it is adjudged worthless (Trow City Directory Co. 1878, p. iv).

The "perfection" of the city directory depended upon a clearly defined system of street addressing, which is why directory publishers often called upon city officials to establish and maintain house numbering systems or went about developing such systems themselves. Yet, even in cities that had adopted house numbering by the mid-19th century, it was a major challenge to develop a house numbering system that could keep pace with the tumultuousness of urban development.

As the urban built environment changed over time, there was a continuous need for city governments to re-number buildings along entire streets to account for such changes. One solution to this issue was the block decimal system, which was developed in Philadelphia in 1856 (Morbito 1965). This system of house numbering allocated 100 numbers for each city block, so that when a new building was constructed or an old building torn down, it would only affect the house numbers on a single block rather than an entire street. Another technique that was developed to address this issue of re-numbering was the equal interval method, which assigned numbers to *street frontage* at regular intervals rather than to specific buildings. The equal interval method was a major departure from previous house numbering practices, because it enabled street addresses to be assigned to parcels that did not have structures built on them and maintained the numbering sequence when new buildings were constructed or old buildings demolished. Both the block decimal system and equal interval method were spatial "solutions" to what Foucault (2007, p. 20) calls the "problem of the series", since they provided a framework for managing the "indefinite series of accumulating units" of buildings by imposing a seemingly permanent spatial order upon the ever-changing urban landscape.

The techniques of house numbering and street addressing that were developed in the 18th and 19th centuries became part of the basic toolkit of modern city planning and municipal engineering during the 20th century. However, street addressing systems were largely confined to urban centers and rarely extended to rural areas. During the 1890s, postal authorities in the United States began home mail delivery to rural residents, which led to the creation of rural route and box numbering systems to identify mailboxes along mail routes (Fuller 1964). These systems of rural addressing served the needs of postal delivery for much of the 20th century. However, there were growing calls by emergency managers to replace rural route and box numbers with city-style street addresses during the 1980s and 1990s (Gribb et al. 1990; Curry et al. 2004; Rose-Redwood 2012a), and similar rural addressing systems have been adopted in other countries such as France as well. During the later 20th century, these changes occurred as emergency response systems were being upgraded with automatic number identification (ANI, commonly known as "Caller ID") as well as automatic location identification (ALI) capabilities. The latter automatically links a street address with each incoming 911 emergency call, but it requires address information formatted with a unique house number and street name for each location (Curry et al. 2004).

The implementation of "Enhanced 911" databases resulted in a major transformation of rural addressing practices across the United States and beyond. The readdressing of the countryside with city-type house numbering and street naming systems was a strategy to produce spaces of security as part of the urbanization of the territory. The security implications of implementing a system of "911 addressing" became particularly evident following the terrorist attacks of September 11, 2001, after which the production of the national territory as a space of geo-locational legibility became a governmental priority. In some cases, rural residents in places such as West Virginia resisted what they perceived as the government's intrusion on private property rights, yet 911 addressing has now been extended throughout many rural communities in the United States (Rose-Redwood 2012a).

Globally, street addressing has been adopted in numerous countries around the world. In China, modern street addressing was likely first introduced in the mid-19th century, yet street addressing systems only became widespread during the second half of the 20th century (Road Naming and Building Numbering Support Group 2005). The generic designations for streets and roads generally consist of two names ( , lù, designating historical streets in a north–south direction, and , jiē, designating commercial streets in an east–west direction). By contrast, the "street" is not the primary spatial unit of addressing in Japan, which instead bases its addressing system upon the district (Cho or Machi , Chome 丁 ), city block (gaiku , ban ), land-lot (banchi 地) and building number (bango 号). With the exception of Higasine-shi, Kyoto and Urakawa-gun, most Japanese cities and towns do not use street names to designate addresses (Rose-Redwood and Kim 2020).

Following the end of the Korean War (1950–1953), North Korea implemented a street addressing system with many streets (거리, geori) named in honor of the ruling family and celebrating the socialist revolution (Kim 2014). South Korea, on the other hand, has long used its traditional land-lot numbering system ( 住 , Jibeon Juso) and only converted to a house number/street name-based addressing system nationwide in 2014, which caused considerable disruption across the country. South Korean addresses are now designated by street name, building number, room number, etc., with roads distinguished by width: Daero (大 , over eight lanes), Ro ( , two to seven lanes) and Gil (길, others). Some South Korean street names have even been commercialized – with corporations acquiring naming rights (e.g. Samsung-ro, LG-ro, SK-ro, etc.). This new South Korean street addressing system has encountered various forms of resistance, yet the government has continued to roll out the new system despite these challenges (Kim 2016).

In sub-Saharan Africa, street addressing plans were imposed in select areas during the colonial era. Following independence from colonial rule, numerous streets in postcolonial cities such as Nairobi were renamed as part of the process of decolonization (Wanjiru and Matsubara 2017). Yet, many African countries still did not have street addressing systems in most informal settlements beyond the formal urban districts in city centers by the turn of the 21st century (Bigon and Njoh 2018). During the mid-1980s, the World Bank and French Ministry of Foreign Affairs began promoting and financing the use of street addressing as part of the "toolbox" of international development (Farvacque-Vitkovic et al. 2005, p. viii). Between 1989 and 1994, house numbering campaigns were launched in many of the larger towns in over a dozen sub-Saharan African countries (Farvacque-Vitkovic et al. 2005). These numbering systems were based on a decametric plan devised by French architect Jean-Jacques Huvé as early as 1805, where house numbers are assigned based upon the distance of a house from a defined zero point, with odd and even numbers used on opposite sides of the street (Pronteau 1966).

The adoption of street addressing systems has been used as a spatial strategy to facilitate municipal tax collection in contexts where cadastral property mapping is not readily available. As a World Bank report explains, "street addressing certainly improves tax collection, but this is probably not a very good advertising slogan" (Farvacque-Vitkovic et al. 2005, p. 46). Instead, World Bank officials have framed street addressing in terms of the "broader recognition of the civic rights of all citizens" (Farvacque-Vitkovic et al. 2005, p. xi). Along with cadastral mapping, street addressing also aligns with the World Bank's neoliberal policies that have encouraged the formalization of informal settlements as a means of establishing officially recognized private property rights in order to extend capitalist economic relations globally.

## 5.4. The future of street addressing and the making of a geocoded world

Critical scholarship on the spatial history of street addressing is only beginning to trace the historico-geographical trajectories, processes and practices that gave birth to the modern street address. In this chapter, we have sought to offer a genealogical account of both the scholarly literature on, and historical emergence of, house numbering and street addressing as political technologies of socio-spatial ordering. Much of the existing scholarship on the history of house numbering has taken its inspiration from Foucault's (2007) theorization of governmentality with its emphasis on the "minuteness of the details" (1995, p. 213) of how power is exercised through the making of spatial regimes of visibility. As the present chapter clearly illustrates, the numbering of houses and naming of streets have played a crucial role in the spatial ordering of cities and towns since the 18th century and continue to shape the urbanization of the territory up to the present.

Although recent studies have greatly improved our understanding of the historical geographies of house numbering and street addressing, there are numerous avenues of potential research that deserve further attention. For starters, the majority of the current literature has focused primarily on the historical development of street addressing in Europe and North America (yet, see Farvacque-Vitkovic et al. 2005; Bigon 2009; Harris and Lewis 2012; Bigon and Njoh 2018; Mask 2020). More research in other world regions is therefore needed if we are to provide a truly "global" genealogy of the street address. Such works could further investigate the influence (or lack thereof) of European colonial rule on the uneven development of street addressing historically as well as the increasing influence of international development agencies in promoting street addressing campaigns in the Global South. Likewise, there is a need to move beyond the urban-centric focus of much of the existing literature by also considering how the political technologies of street addressing are reshaping rural geographies as well.

In methodological terms, historical studies of street addressing have largely relied upon textual sources found in archives. Archival resources are, of course, indispensable to genealogical accounts of the house number and street address, but other methodological techniques such as oral histories, interviews, participant observation and digital ethnographies can supplement archival sources, especially for histories of street addressing that are still unfolding in the contemporary era. Such methods are better equipped to examine the ways in which the street address has become incorporated into the practices of everyday life through the taken-for-granted rituals of habit and routine (Light and Young 2014).

There are also various thematic areas of research that require further critical analysis. In particular, the lack of city-type street addressing has recently been used as the basis for political disenfranchisement in the United States, where some rural residents and members of Indigenous communities have been barred from voting because they do not have a residential street address (Mask 2018). A history has yet to be written on the use of street addressing as a political technology of voter suppression. Yet, street addressing has also been a force of political enfranchisement, such as when the dehumanizing "Native Yard" street numbering system of apartheid-era South Africa was transformed into a spatial framework linking addressing and citizenship during the democratization process (Giraut and Houssay-Holzschuch 2016).

Additionally, the intersections between the genealogies of street addressing and the digital geographies of geocoding have the potential to become a significant research focus of both spatial history and critical digital studies (Sotoudehnia 2018). Some private companies, such as what3words (whose motto is "Addressing the World"), have even sought to bypass street addressing by developing commodified digital geocoding platforms that do not directly rely upon house numbering and street naming systems. Yet, most digital geo-navigation systems still rely extensively on street address databases, since latitude/longitude coordinates or what3words geocodes on their own do not provide a means of determining the best routes to take on the ground between one location and another.

These are just a few of the potential topics that might be explored by future scholarship on street addressing and the production of geocoded worlds. There is still much we do not know about the histories of street addressing, yet the addressing of the world is not a matter of history alone. It is also part of an ongoing process of uneven geographical development and state formation, and tracing the political genealogies of the street address can therefore help us better understand the making of the spatial histories not only of the past but of the present and future as well.

## 5.5. References


Alotta, R. (1990). *Mermaids, Monasteries, Cherokees and Custer: The Stories Behind Philadelphia Street Names*. Bonus Books, Santa Monica.

Berg, L. and Vuolteenaho, J. (eds) (2009). *Critical Toponymies: The Contested Politics of Place Naming*. Ashgate, Farnham.

Bigon, L. (2009). Urban planning, colonial doctrines and street naming in French Dakar and British Lagos, c. 1850–1930. *Urban History*, 36(3), 426–448.

Bigon, L. and Njoh, A. (2018). Toponymic complexities in sub-Saharan African cities: Informative and symbolic aspects from past to present. In *The Political Life of Urban Streetscapes: Naming, Politics, and Place*, Rose-Redwood, R., Alderman, D., Azaryahu, M. (eds.). Routledge, New York.

Cicchini, M. (2012). A new "Inquisition"? Police reform, urban transparency and house numbering in eighteenth-century Geneva. *Urban History*, 39(4), 614–623.

Cicchini, M. (2018). Les incertitudes du commencement. La numérotation des maisons et sa réception à Genève à la fin du XVIIIe siècle. *Histoire urbaine*, 3(53), 107–125.

Coleman, J. (2016). The families whose homes were demolished by mistake. *BBC News*, March 26 [Online]. Available at: https://www.bbc.com/news/magazine-35897095 [Accessed 17 March 2020].

Curry, M., Phillips, D., Regan, P. (2004). Emergency response systems and the creeping legibility of people and places. *The Information Society: An International Journal*, 20(5), 357–369.

Farvacque-Vitkovic, C., Godin, L., Leroux, H., Verdet, F., Chavez, R. (2005). *Street Addressing and the Management of Cities*. World Bank, Washington, DC.

Foucault, M. (1995). *Discipline and Punish: The Birth of the Prison*, translated by Alan Sheridan. Vintage Books, New York.

Foucault, M. (2007). *Security, Territory, Population: Lectures at the Collège de France, 1977–1978*, translated by Graham Burchell. Palgrave MacMillan, New York.

Fuller, W. (1964). *RFD: The Changing Face of Rural America*. Indiana University Press, Bloomington.

Garrioch, D. (1994). House names, shop signs and social organization in western European cities, 1500–1900. *Urban History*, 21(1), 20–48.

Garrioch, D. (2002). *The Making of Revolutionary Paris*. University of California Press, Berkeley.

Giraut, F. and Houssay-Holzschuch, M. (2016). Place naming as *dispositif*: Toward a theoretical framework. *Geopolitics*, 21(1), 1–21.

Goodman, S. (1954). *A Bibliography on the Subjects of House Numbering, Street Numbering, Street Names, and Street Name Signs*. Goodman, Long Island City.

Gribb, W.C. and Harrington Jr., J. (1990). Rural addressing and computer mapping in New Mexico. *The Professional Geographer*, 42(4), 471–480.

Harris, R. and Lewis, R. (2012). Numbers didn't count: The streets of colonial Bombay and Calcutta. *Urban History*, 39(4), 639–658.

Heal, A. (1942). The numbering of houses in London streets. *Notes & Queries*, ClCCCIII, 100–101.

Henkin, D. (1998). *City Reading: Written Words and Public Spaces in Antebellum New York*. Columbia University Press, New York.

Hunt, N. (1995). House numbering in revolutionary Paris. *Bulletin of the Institute of Mathematics and Its Applications*, 31, 9–10.

Joyce, P. (2003). *The Rule of Freedom: Liberalism and the Modern City*. Verso, London.

Kim, K.-H. (2014). Changes of urban structure of Pyongyang city represented in road names. *Journal of Cultural and Historical Geography*, 26(3), 34–55 (Korean, abstract in English).

Kim, S.-B. (2016). A study on the named sources of the Korean street naming: Focusing on the centripetal and anthroponym street names. *Journal of the Place Name Society of Korea*, 25, 33–93 (Korean, abstract in English).

Light, D. and Young, C. (2014). Habit, memory and the persistence of socialist-era street names in post-socialist Bucharest, Romania. *Annals of the Association of American Geographers*, 104(3), 668–685.

Magnusson, W. (2011). *Politics of Urbanism: Seeing Like a City*. Routledge, Abingdon.

Marx, K. (1990). *Capital: A Critique of Political Economy*, translated by Ben Fowkes. Penguin, New York.

Mask, D. (2018). Where the streets have no names, the people have no vote. *New York Times*, October 19 [Online]. Available at: https://www.nytimes.com/2018/10/19/opinion/sunday/north-dakota-addresses-voting-id.html [Accessed 17 March 2020].

Mask, D. (2020). *The Address Book: What Street Addresses Reveal about Identity, Race, Wealth, and Power*. St. Martin's Press, New York.

Morbito, J. (1965). Physical and social implications of street naming and house numbering systems in today's urban environment. Unpublished Master's Thesis, School of Urban Planning and Landscape Architecture, Michigan State University.

Njoh, A. (2010). Toponymic inscription, physical addressing and the challenge of urban management in an era of globalization in Cameroon. *Habitat International*, 34(4), 427–435.

Pronteau, J. (1966). *Les Numérotages des maisons de Paris du XVe siècle à nos jours.* Publications de la sous-commission de recherches d'histoire municipale contemporaine VIII, Paris.

Road Naming and Building Numbering Support Group (2005). World address system. Report, South Korea's Ministry of Government Administration and Home Affairs, Seoul (Korean).

Rose-Redwood, R. (2006). Governmentality, geography, and the geo-coded world. *Progress in Human Geography*, 30(4), 469–486.

Rose-Redwood, R. (2008). Indexing the great ledger of the community: Urban house numbering, city directories, and the production of spatial legibility. *Journal of Historical Geography*, 34(2), 286–310.

Rose-Redwood, R. (2012a). With numbers in place: Security, territory, and the production of calculable space. *Annals of the American Association of Geographers*, 102(2), 295–319.

Rose-Redwood, R. (2012b). A regular state of beautiful confusion: Governing by numbers and the contradictions of calculable space in New York City. *Urban History*, 39(4), 624–638.

Rose-Redwood, R. and Kadonaga, L. (2016). "The corner of Avenue A and Twenty-Third Street": Geographies of street numbering in the United States. *The Professional Geographer*, 68(1), 39–52.

Rose-Redwood, R. and Kim, S.-B. (2020). Street naming and power. In *International Enclopedia of Human Geography*, 2nd edition, Kobayashi, A. (ed.). Elsevier, Amsterdam.

Rose-Redwood, R. and Tantner, A. (eds) (2012). Governmentality, house numbering, and the spatial history of the modern city. *Urban History*, 39(4), 607–613.

Rose-Redwood, R., Alderman, D., Azaryahu, M. (eds) (2018). *The Political Life of Urban Streetscapes: Naming, Politics, and Place.* Routledge, London.

Sasse, C. (2021). Die Stadt lesen. Englische "Directories" als Wissens- und Orientierungsmedien, 1760-1830. De Gruyter/Oldenbourg, Berlin and Boston.

Smail, D. (2000). *Imaginary Cartographies: Possession and Identity in Late Medieval Marseille.* Cornell University Press, Ithaca.

Sotoudehnia, M. (2018). Toponymic checksum or flotsam? Recalculating Dubai's grid with Makani, the smartest map in the world. In *The Political Life of Urban Streetscapes: Naming, Politics, and Place*, Rose-Redwood, R., Alderman, D., Azaryahu, M. (eds). Routledge, New York.

Tantner, A. (2007). *Ordnung der Häuser, Beschreibung der Seelen – Hausnummerierung und Seelenkonskription in der Habsburgermonarchie.* Studienverlag, Innsbruck.

Tantner, A. (2009). Addressing the houses: The introduction of house numbering in Europe. *Histoire & Mesure*, 24(2), 7–30.

Tantner, A. (2015). *House Numbers: Pictures of a Forgotten History*. Reaktion Books, London.

Thale, C. (1986). Milwaukee's street numbering system. *Milwaukee History*, 9(3), 66–76.

Thale, C. (2007). Changing addresses: Social conflict, civic culture, and the politics of house numbering reform in Milwaukee, 1913–1931. *Journal of Historical Geography*, 33(1), 125–143.

Trow City Directory Co. (1878). *Trow's New York City Directory*. Trow City Directory Co., New York.

Vuolteenaho, J. (2012). Numbering the streetscape: Mapping the spatial history of numerical street names in Europe. *Urban History*, 39(4), 659–679.

Wanjiru, M. and Matsubara, K. (2017). Street toponymy and the decolonisation of the landscape in post-colonial Nairobi. *Journal of Cultural Geography*, 34(1), 1–23.

Wittstock, B. (2008). *Die Berliner Hausnummer. Von den Anfängen Ende des 18. Jahrhunderts bis zur Gegenwart im deutschen und europäischen Kontext*, 5 vols. Pro Business Verlag, Berlin.

# 6

# Toponymic Commodification: Thematic Brandscapes, Spatial Naming Rights and the Property–Name Nexus

**Jani VUOLTEENAHO**
*Faculty of Social Sciences, University of Helsinki, Finland*

## 6.1. Introduction

> In order to better understand the process of producing toponymy-as-commodity, we must delve more deeply into the interrelations of naming, property, forms of capital and rent-seeking. (Rose-Redwood et al. 2019a, p. 849)

Increasingly treated by economic and political elites as a set of techniques and tokens to upgrade place images, encourage inward investment, or directly add to property owners' cash flows, toponymy is one of the commodification frontiers of the neoliberal era (Vuolteenaho and Ainiala 2009; Rose-Redwood et al. 2010, p. 466; Madden 2019; see Fraser 2014). Literature reviews and case studies by critical toponymists show that the processes of toponymic commodification are underway in many guises (e.g. Berg 2011; Rose-Redwood and Alderman 2011; Light 2014; Medway and Warnaby 2014; Karimi 2016; Wideman and Masuda 2018; Sotoudehnia 2018; Spocter 2018; Crețan 2019; Kearns and Lewis 2019; McElroy 2019). This chapter takes two major breeds of "toponymy-as-commodity" (Rose-Redwood and Alderman 2011; Light and Young 2015) under closer scrutiny:

*The Politics of Place Naming*,
coordinated by Frédéric GIRAUT and Myriam HOUSSAY-HOLZSCHUCH. 

commercially motivated uses of *thematic place naming* (an integral part of promotional spatial projects both presently and in earlier phases of capitalist development), and the selling of *spatial naming rights* (a recently globalized ingredient of neoliberal policies with only rare if any pre-1990s occurrences outside the US context). Taking its cue from recent calls to conceptualize the commodification contexts of place naming through their entanglements with property relations (Giraut and Houssay-Holzschuch 2016; Madden 2019; Rose-Redwood et al. 2019a), I meanwhile coin the concept of "property–name nexus" as a lens to unravel these two variations of profit-driven naming in a range of North American, European and Asian settings studied below.

Especially the category of thematic place promotion or branding through toponymy (e.g. Raento and Douglass 2001; Vuolteenaho and Ainiala 2009; Vuolteenaho and Kolamo 2012) is a variegated, porous yet also perpetually emergent domain of toponymic commodification. The applications of this technique cover place-based franchises and corporate "brandlands" (from Disneyfication in the strict sense to Legolands, Guggenheim or Louvre museums, Niketowns and BMW Worlds), "Silicon"-, "Smart-" and "Media"-epipheted spaces of high-tech and media industries, landscapes of spatio-temporal "elsewhereness" (epitomized by Dubai's The World archipelago, or Tom Sawyer Islands as Disneyland attractions), sites of heritage revalorization (like Manchester's Castlefield Urban Heritage Park), indoor and outdoor "tropical" paradises, ethnically or artistically branded gentrified neighborhoods, tropes of idyllic "nature" archetypal in suburban toponymies and many more (e.g. Hopkins 1990; Gottdiener 2001; von Borries 2004; Yakhlef 2004; Leary 2009; Johansson and Cornebise 2010; Berg 2011; Medway and Warnaby 2014; Wideman and Masuda 2018; Madden 2018). Both pre-neoliberal and neoliberal-era implementations of such thematic brandscapes, in which the instituting of appealing toponymy on a "place property" by its owner plays a role, are legion.

By contrast, as a genuinely neoliberal domain spatial naming rights pivot around negotiated, exclusive contracts that concern the leasing of a place name per se (Rose-Redwood et al. 2019b; Vuolteenaho et al. 2019a). While interpretations of the exact outset of the phenomenon vary, Rose-Redwood et al. (2019a, p. 848) mention a 25-year leasehold of the naming rights for the Buffalo Bills' stadium to the Rich Products Corporation in 1972 as a path-setting deal that "continues to have significant reverberations today". Especially since the 1990s, spatial naming rights have been on property owners' and municipal policy-makers' lips as an "innovative" additional or non-tax revenue source. All around the world, thousands of naming rights for diverse properties have been thereupon contracted, re-contracted, expired and also contested. The spectrum of privately or publicly owned property types for which naming rights have been allocated include state-of-the-art event venues (Allianz Stadium in Turin, Italy), historic stadiums (United Airlines Field at the Los Angeles Memorial Coliseum), major urban infrastructures (Zhou Hei Ya and Jianghan Road station in Wuhan, China) and even relatively minor city assets (Akman/Shindleman children's program room in the Henderson Public Library, Winnipeg, Canada).

The following comparisons of thematic place naming and spatial naming rights, and their embeddedness in specific spatial projects and property relations, make it plain that the practices of toponymic commodification are not only reshaping the particularities of the place, but also convey cross-contextual patterns, processes and power hierarchies (see Peck 2015). Theoretically, the rationale throughout the chapter is to bring recent openings into toponymic commodification by critical place naming scholars into a fuller dialogue with relevant "non-toponymy-centric" theorizations of the commodification of space, place and landed property (e.g. Lefebvre 1991; Blomley 2004; Logan and Molotch 2007). Critical toponymists keen on commodification have productively drawn from the theories of economic and symbolic capital, and discussions around neoliberal urbanization and its policy mobilities aspects (e.g. Bourdieu 1986; Brenner and Theodore 2002; McCann 2013; Lauermann 2016; Giraut and Houssay-Holzschuch 2016; Madden 2019; Rose-Redwood et al. 2019b). In this chapter, I seek to enrich this repertoire through explicit insights into property relations and rent-garnering by theorists in the political economy lineage (e.g. Polanyi 1944–2001; Harvey 1973, 2012; Rose 1994; Christophers 2010; 2018; Haila 2016; Penny 2016). Specifically, I will apply the critical theorizations of exclusive property rights (incubating the commodified forms of thematic naming and spatial naming rights, yet colliding with other types of non-commodified spatial rights), the differentiation of rental values, relations between privatization, publicness and regulation, and contractualization (used also in toponymic contexts as a way of outsourcing of public governance) to investigate intricate power relations behind the place- and time-specific manifestations of toponymic commodification.

I commence the chapter's mainly literature-based excavations with four case-specific illuminations of thematic branding through toponymy. Three of these are large-scale master-planned real estate developments acted out in varied geohistorical settings, one a gentrification-related regeneration process: the first case highlights a pre-neoliberal spatial project epitomizing the owner's exclusive property rights (the inaugural Sun City new town in Arizona), the second a waterfront development based on public–private partnering (the Aurinkolahti development in suburban Helsinki), the third a prestigious South-East Asian development on a "post-socialist" frontier of real estate capitalism (the Vinhome Central Park in Ho Chi Minh City), and the last one a toponymic gentrification process in Brooklyn's Dumbo area. Next, I study spatial naming rights as an ownership-grounded, in essence contract-based, form of name-making. In here, the key specimens are taken from cities like Philadelphia (NRG Station, formerly Pattison Station and AT&T Station), Dubai (naming rights campaign for metro stations), Winnipeg (an extensive program to sell naming rights for city assets), Turku (Gatorade Center as the fourth consecutive sponsored name for an indoor arena), Newcastle (time-honored St. James' Park stadium with two short-lived corporate names) and New York City (Barclays Center indoor arena and adjacent Atlantic Avenue–Barclays Center subway station). With a focus on the renaming of publicly owned properties, I draw from these and other vignettes to discuss naming rights as the emblems of quasi-privatized public governance, as the objects of contestation, and as manifestations of gaping rental value differences that typify the spatial naming rights marketplace. In conclusion, I elaborate on the concept of property–name nexus – encompassing both straightforward instances where those who own a property play a deciding role in its naming (common in the production of thematic brandscapes), and more complicated cases of outsourcing and contracting out an owner's right to name (typical for spatial naming rights) – as a critical window on the workings of toponymy as a neoliberal commodification frontier. In distinct ways, both the thematic and naming rights-based techniques of toponymic commodification hinge on the capitalist, state-sanctioned doctrine of the owner's exclusive property rights, and contribute to the power-infused and profit-driven productions of space in particular socio-cultural and political-economic contexts.

## 6.2. Thematic namescapes in branding neighborhoods: From Sun Cities to Icebar Saigon and a Brooklyn with distinction

With a long history as an instrument of place boosterism, toponymic practices distinctive to the so-called themed economy (Gottdiener 2001) continue to thrive in contemporary cities and regions (Sorkin 1992; Raento and Douglass 2001; Miles 2010, pp. 142–163; Medway and Warnaby 2014; Zhao et al. 2019; Caiazzo et al. 2020). Thematic place naming often bears the stamp of a "culture of the copy" (Klingmann 2007). Raento and Douglass (2001) classified naming motifs across the Las Vegas Strip into ones that connote (1) good fortune, (2) wealth and opulence, (3) adventure, excitement and fantasy, (4) geographical references, (5) temporal evocations, (6) intimacy or informality and (7) names that incorporate "power words" of corporate branding. In a reminiscent analysis in the Finnish context, Vuolteenaho and Kolamo (2012) studied five categories of appealing generic motifs that array entertainment and high tech landscapes, including the themes of (1) spatio-temporal escapism, (2) power, money, centrality or grandeur, (3) speed or technological advancement, (4) positively valued “universals” such as happiness, good fortune and sunshine, and (5) names with a global linguistic prestige (typically borrowings from English and to a lesser degree from Latin, French and Italian vocabularies). These taxonomies imply that the profit-oriented uses of toponymy often involve an emulative, simulative or extrospective logic (Soja 1989, pp. 244–245; Gottdiener 2001; see also McCann 2013), and above all speculate on a place name’s economic value (Giraut and Houssay-Holzschuch 2016, pp. 9–10) to persuade wanted mass or niche market audiences.

### 6.2.1. *A pre-neoliberal piece of toponymic place branding: the housing and leisure idyll of Sun City, Arizona*

As an entrepreneur-led project from the ground up, our first vignette is the outcome of a very straightforward property–name nexus imposed on a former cotton farm by the Del E. Webb Development Company (DEVCO), a real estate titan otherwise known for its involvement in building mega-size “theme hotels” in Las Vegas (e.g. the legendary Flamingo Hotel opened in 1946 and the Sahara Hotel opened in 1952), and dozens of other large-scale constructions across the US. Our case is the inaugural Sun City “new town” in the outskirts of Phoenix, catering exclusively to people 55 and older. The legendary enclave for “active” older-age living was opened in 1960, a half year after DEVCO had purchased a lot that suited its business vision from J.G. Boswell Company (Findlay 1992, pp. 174–175). As an instance of the surge of mass-produced suburbia in the post-WWII US context, DEVCO drew from the techniques of “selling the suburb” pioneered already in the late 19th century (Ward 1998, pp. 83–140). Also in Sun City’s case, the marketed product was an “exclusive and permanent rights to a small part of the whole suburb together with non-exclusive rights to enjoy the place in its totality” (Ward 1998, p. 84). As detailed by John M. Findlay (1992), the originality of Sun City’s concept was in marketing this exclusively to senior citizens. In a new way, the aging American population was given the promise of “partaking in the joy of living on their own terms” within a “nearly self-sufficient” landscape with modern amenities and recreations. DEVCO rendered the acquisition of a Sun City retirement home a simple purchasable package. On-site toponymic odes to sun(shine), greenness and leisure opportunities emblazed a mass-produced place brand first of its kind.

As an admixture of leisure-oriented infrastructure and allusively themed toponymy, the development was planted as a carefully imagineered reverse of urban seediness (see also Beauregard 2006) – as a “country club” type of environment (Findlay 1992). Within the development, all was designed with the greatest care to warrant this image. The layout harbored artificial ponds, golf courses and "green acres" that effectively insulated its overwhelmingly white lower middle-class/middle-class retiree population from the reality of the surrounding desert (Findlay 1992, p. 203). In tune with this, many of Sun City's roads were named after "nationally famous golfers and golf courses" – North Augusta Drive, Cherry Hills Drive, West Brookside Drive and more fell into this odonymic sub-theme. Both these and the preponderance of other street toponyms that still striate Sun City recycle the tropes of "nature" and "pastoral gentry" as the "banal" perennials of the theming of suburban idylls in the US and beyond (Norris 1999; Wood 2002; Berg 2011). Crowning the place product, the "sun" in the spatial brand name itself self-explanatorily evoked Arizona's warm climate, and particularly "a chance to get away from colder climates" (Findlay 1992, p. 196), a message that DEVCO's "formidable marketing machinery" efficiently disseminated in years to come to catch up its niche nationwide (Findlay 1992, p. 176)[1].

### 6.2.2. *Public–private partnering in Helsinki: revamping the "Sun Bay" suburb toponymically*

Our next episode of thematic place branding through toponymy concerns Aurinkolahti ("Sun Bay") neighborhood in suburban Helsinki, studied previously by me and my colleagues from toponymic perspectives (Vuolteenaho and Ainiala 2009; Vuolteenaho et al. 2019b), and by the land rent theorist Anne Haila (2005) from a development rights perspective. Unlike with the Sun City as a single-handed corporate creation, Aurinkolahti's rise into an upmarket, flamboyantly named seaside housing landscape was a complex result of public–private partnering, propelled by an entrepreneurial turn in the City of Helsinki's policies applied to the area. In Finnish legislation, urban planning from the master planning process to the minutiae of street naming remains the prerogative of local authorities (see on planning systems in the US, European and East/South-East Asian settings: Haila 1999). In Aurinkolahti's case, however, the City suspended its "planning monopoly" in favor of ProPaulig development company, an affiliate of a local coffee roaster and seasoning manufacturer Paulig. Largely due to failures in preceding development projects in Aurinkolahti's proximity, the City Executive Office found ProPaulig's

1. As Pacione (2012, p. 165) encapsulates further phases in this piece of American corporate history, in decades to come DEVCO developed 10 master-planned Sun Cities (from Arizona to South Carolina) and tens of other retirement communities, "comprising 80,000 homes in which age restrictions, security patrols and gates and walls create[d] enclaves of nearly self-sufficient 'age specific' housing". Sixty years since launching the concept, Del Webb is still operating in the "active adult" real estate business as a brand of PulseGroup (www.delwebb.com).

blueprint (inspired by vogueish international exemplars of revitalized waterfronts) for the area's industrialized and hitherto unbuilt coastal swathes as endorsable. As an instance of neoliberal outsourcing of public assets and governance (Penny 2016), the city sold its key holdings with favorable covenants to ProPaulig, and authorized the latter to take lead in Aurinkolahti's design, implementation and negotiations with private investors, while itself taking a role as a partner who directed taxpayer funding in the area's infrastructure and marketing (Haila 2005). Exceptionally, the contrivance of the area's street toponymy was also contracted out to private developers (Vuolteenaho and Ainiala 2009).

All this fed into a commodified, self-assertive thematic toponymy that stood in a marked contrast to traditionalist and prosaic suburban nomenclatures favored by planners elsewhere in Helsinki and Finland (Vuolteenaho and Ainiala 2009, p. 231, 246). Around the mid-1990s, the whole area's name was reverted from a "somber" Mustalahti ("Black bay") to "sunshiny" Aurinkolahti – an unlikely appellation for a northern European suburb – to sell the suburb to investors, developers and affluent home-seekers. Also otherwise, a multi-layered street and condominium neotoponymy did it damnedest to revamp Aurinkolahti as a place "very close to wonderland" (ibid., 241–247). For one thematic focus, local heritage embedded in the area's upper-class villas (*Ivan Falin's Lane, Midnight Summer Rose Lane*)[2] was put into the service of "the needs of the here and the now" (Soja 1989, pp. 244–245). Given the branch of industry of the area's decades-long patron, coinages like *Coffee Street, Seville Orange Park, Bertha Paulig Street, Helsinki's Cappuccino* and *Helsinki's Corretto* were at once retro-spirited and evocative of trendy urban culinary culture. In more escapist fashion, Aurinkolahti's toponymy also incorporated evocations of the touristic trinity of "sun", "sea" and "sand" (*Sun Beach, Gold Sand Square, Hot Spell Square, Park of Beautiful Weather, Helsinki's Sun Gold*, and many other names with "summerly" overtones), geographical or historical "elsewhereness" (*Columbus* shopping mall, *Helsinki's Gondolier*) and sea-related derivations from ancient Greek and Roman mythology (*Helsinki's Aphrodite, Helsinki's Poseidon, Helsinki's Triton*). Unabashedly catered to varying niches of well-to-do home-seekers, some of the area's newcomers interviewed by our research team saw no problems in such toponymic flamboyancy within a "really fabulous" new neighborhood (Vuolteenaho et al. 2019b). By contrast, not few unaccustomed residents in surrounding areas were irritated by Aurinkolahti's aggressively market-oriented promotion, opining that "many of its place names sound horribly trumped-up" (Tegelman 2006).

2. All street and condo names mentioned in this subsection are translations from Finnish.

### 6.2.3. *A South-East Asian real estate hotspot read toponymically: Vinhomes Central Park*

Announced in 2014 as a USD 1.6 billion waterfront renewal, our South-East Asian vignette of commercialized thematic neotoponymy focuses on the *Vinhomes Tân Cảng* (in Vietnamese) or *Vinhomes Central Park* (VHCP), billed by its marketers as "the jewel in Ho Chi Minh City crown", and by *Vietnam Investment Review* as "a benchmark property in the city"[3]. The luxury development's rapid completion is a testimony to the rise of Vietnam into the latest high-yielding real estate hotspot in South-East Asia, accompanied with blatantly westernized and above all Anglicized toponymy (see information on post-socialist Russia: Yurchak 2000). As critics like Erik Harms (2012) and Tim Doling (2019) note, Ho Chi Minh City's (HCMC) government has very actively attracted large-scale inward investments through the destruction of historic structures, programmatic evictions, liberalized housing ownership regulations and speculative land policies, legitimized as "urban beautification". In particular, the local authorities have allowed the Vingroup corporation and its affiliate Vinhomes[4], the biggest residential property developer by far in Vietnam, "to buy up huge swaths of prime riverfront property" (Harms 2019, p. 64). The walled 44-hectare estate of VHCP is located in one of these plots. Kept at bay from the chaotic street life in the rest of HCMC, the key ingredients planted on the enclave include a spacious, immaculate *Central Park* (lending its name to the whole area), a cluster of high-end villa apartments, a score of residential skyscrapers, and the 81-storey Vincom Landmark 81 (the world's 17th tallest skyscraper). Made possible by the city's neoliberalized land policies, the colossal enclave is through and through a tour de force by its corporate owner Vinhomes (in this regard reminiscent of DEVCO's Sun City six decades earlier).

Carrying out fieldwork in VHCP in November 2019, it was difficult to escape an impression of a gargantuan "theme village" under corporate tutelage. Two thematic aspects of place branding had an overwhelming presence in the on-site visual namescape. First and foremost, self-iterative appellations associated with the owning brand chain's Vin-affixed "power" words and compounds (Raento and Douglass 2001) were everywhere in the corporate new town: from *Hotel Vinpearl Luxury Landmark 81*, *Vincom Center Landmark 81 Shopping Mall* and *Vincom Ice Rink* to the signage of *VinSchool*, *VinMed* hospital, *VinMart* groceries and other amenity

3. https://www.vir.com.vn/vinhomes-project-central-to-ho-chi-minh-citys-plan-39526.html.

4. Besides *Vinhomes Central Park*, Vinhomes is the developer of around forty mega-developments across Vietnam, including *Vinhomes Star City*, *Vinhomes Smart City*, *Vinhomes Ocean Park*, and many more estates that bear its brand name. Both Vinhomes and its parent conglomeration Vingroup are in the majority ownership of the Vietnamese billionaire Phạm Nhật Vượng.

suppliers interspersed throughout the development. Second, the area's housing nomenclature was based on the reiterations of the motifs of "Landmark" (evocative of grandeur and sightliness) and "Central Park" (an evident allusion to Manhattan's world-famous Central Park). Two separate clusters of its residential towers were named and sub-numbered as Parks 1–7 and Centrals 1–7, the rest of them as Landmarks 1–6 and Landmark Plus (the latter with serviced apartments). As with these "presentation names" (Chen 2017), the *Landmark 81*'s key touristic or recreational attractions – including *Skytouch* parachute jumping-simulator, *Skyview Observation Deck* and *Icebar Saigon* (touted as "the highest ice bar in the world" – also bore English language names only, as if "urban beautification" were synonymous with Anglicization (see Harms 2012). Usages of native Vietnamese language were prominent on guide boards and small shop signage, but largely absent from major commercial nomenclature. Traces of national or local pasts were even more efficiently marginalized. The only exceptions I was able observe were two public roads that traversed the area and unremarkably commemorated historic Vietnamese notables. The overall appearance of the namescape of *Vinhomes Central Park* was ahistorical and westernized – streamlined around a few well-trodden promotional themes (spectacular tallness, the Central Park motif, exotic "wintriness" amid HCMC's tropical urbanism), highly brand-centric (the ubiquitous salutes to the owner's business empire) and catered to cosmopolitan Vietnamese and expat tastes.

### 6.2.4. *A coda on ownership rights and thematically named brandscapes*

As noted, the profit-driven variations of thematic place-making and -naming are by no means restricted to large-scale master-planned and mainly residential developments as above. The highlighted place- and time-specific mobilizations of thematic toponymy are though sufficient for some tentative synthesizing interpretations. Clearly, both a political-economic context (economic ups and downs, action/inaction by the local state, planning legislature, availability of developable land, etc.) and socio-cultural circumstances (targeted audiences, place-embedded histories, scalar relations to wider local/national/global social space) matter greatly in the practices of toponymic commodification. In all three vignettes, an entrepreneurial actor with to a greater or lesser degree "full" development rights to act on the intervention area – in Sun City and Vinhomes Central Park a heavy-weight real estate tycoon, in Aurinkolahti a local developer in close partnership with city authorities – imposed its brand visions onto the namescape. Emulated and well-trodden motifs (barring certain clues in Aurinkolahti's neotoponymy to local heritage) and symbols of corporate power (particularly salient in VHCP's case) spoke the loudest in these cases of toponymic commodification. Notably, in all cases, preceding landownership arrangements made it possible to reimagine a place and its nomenclature as if ex nihilo.

In more "saturated" urban spaces with multiple overlapping social histories, competing interests and associated claims for the right to the city, however, such a straightforward instituting of a neotoponymy is far from always the case (e.g. Rose 1986; Johansson and Cornebise 2010; Huron 2015; Alderman and Ingwood 2018). Particularly in cities, the acts of naming are the contested acts of claiming and replacing (Blomley 2004). This is graphically illustrated by struggles around the "new toponymies of gentrification" in Brooklyn analyzed by David Madden (2018, p. 1604). In the NYC borough, middle-class "brownstoners" started, as early as the 1960s, to take back forgotten urban names as the tools of neighborhood upgrading and replacing of stigmatized working-class community names. More recent struggles have emerged as the neoliberalized local governance has pushed forward "the emergence of new SoHos and TriBeCas" (Madden 2014, p. 488), while real estate speculators have purchased low-yielding properties to redevelop them into high-yielding luxury areas (Madden 2018, p. 1605). Such a "rent gap reclaiming" occurred in an ex-warehouse area now known as Dumbo. Originally a "dadaist" coinage (from "Down Under the Manhattan Bridge Overpass"), a group of local artists used the Dumbo moniker initially as an anti-gentrification resistance token "to protect our turf from developers" (ibid., 1607–1608). However, forces of corporate-led redevelopment in the lead of Two Trees real estate company "quickly saw the value of such a name as a way to associate the neighborhood brand with art, culture, and the loft-dwelling narrative", and started to buy the area's buildings. Dumbo gradually became a commodified "prestige name", and a vehicle for stigmatizing of extant pockets of affordable housing in its vicinity (Madden 2018). As a kind of trendy umbrella for "a Brooklyn with distinction" (see Trinch and Snajdr 2017), the fittingly enigmatic, "unclichéd" Dumbo appellation became a brand to sell the place and its real estate units, both in situ and online. While Dumbo's regeneration again shows that the political-economic and socio-cultural contexts of toponymic commodification differ dramatically, in its case the process was not based on the invention of a neotoponymy but on the manipulation of a readily named place for its exchange values (see Logan and Molotch 2007).

## 6.3. Buying into and contesting spatial naming rights

As distinct from more conventional forms of toponymic commodification, the selling of spatial naming rights is based on quid pro quo transactions (creating reciprocal responsibilities for the two involved parties) and associated (usually fixed-term) contracts. Law and marketing scholars as well as consultants like the

Chicago-based International Events Group define it as an "exclusive" type of sponsorship in which a fee is paid to property "in return for access to the exploitable commercial potential of that property" (Gillooly and Medway 2018, p. 199). Through such acts of contractual name-making, a toponym is explicitly rendered a legally binding market relation. Irrespective of whether the owner of a place property to which a naming right is allocated is a public or private entity, a signed name leasehold means the intertwinement of the place's and the sponsor's interests in a way that the latter "must now be respected and protected" (Madden 2019, p. 889).

Favorable interpretations of the worldwide naming rights boom during the last three decades tend to emphasize it as an extension to the long-established traditions of stewardship, an innovative financial/fiscal-cum-marketing instrument and a "win-win" vehicle of place boosterism whose positive effects will trickle down to the public (e.g. Burton 2008; Fortunato 2013). In contrast, critics insist that naming rights are fraught with negative externalities and ethical problems (e.g. Boyd 2000; Butterworth and Moska 2009; Rose-Redwood et al. 2019b; Weaver 2019). Through them, a "constant state of flux" is being created "as one corporate sponsor's name replaces another when [...] contracts expire", or stop short for any reason (Rose-Redwood 2019b, p. 748), additional technical-cartographic costs emerge, confusions in everyday cognitions and wayfinding are risked, corporate powers may exert "bad image" laundering, and the conventional functions of toponymy as the carriers of heritage are done away. In the light of these criticisms, one might conceive spatial naming rights as a new toponymic form of "rent-seeking", defined by public choice scholars as "the use of resources to acquire, through political pressure, special privileges whereby the injury to others is greater than the gain to those obtaining rents" (Haila 2016, p. 121; see also Rose 1994; Tullock 2005). In this section, I examine a set of agreements, policies, controversies and power relations related to the selling of naming rights for mainly publicly owned event facilities and transport stations, both of which have spread with amazing speed in recent decades.

### 6.3.1. *Quasi-privatization through and for naming rights sales*

Pronouncedly in our neoliberal times, "public landowners do not behave differently from private landowners at all" (Christophers 2018, p. 339). This is mirrored in the naming rights marketplace, populated not only by private corporations but also governments and authorities in contractual and overseeing roles. Carol Rose (1986, p. 713) noted that in western legal tradition (from Roman Law and to modern American Law) there exists "a curious cross-current" that certain kinds of property – spaces of transport and travel, customary places for communal events, for instance – "should not be held exclusively in private hands", and that public regulation has aimed to "prevent private owners from capturing the 'rent' created by 'publicness'" of such "inherently public properties" (ibid. pp. 770–774). However, neoliberalized local governments have become increasingly oblivious to this principle. This is also occurring in toponymic terms, and particularly through allowing the selling of naming rights for landmarks and hubs of local life, producing the "enclosures" of publicness (Rose 1986; Madden 2019; Weaver 2019).

Timothy Weaver's (2019, pp. 437–440) analysis of the renaming of Pattison station in Philadelphia is a case in point. Owned by the Southeastern Pennsylvania Transportation Authority (SEPTA), the transit hub was renamed AT&T station in 2010 after the telecommunications giant and eight years later as NRG station after the energy company. Adding a rental value to its name, the station functions as a gateway to a major leisure complex housing baseball, football, basketball, hockey megastructures and other event arenas. Weaver contends that the primary benefits of the named leaseholds have flown to the name-lending sponsors via their greatly enhanced public exposure: "Taking public transportation to see a sports team in one of the nation's great cities involves being confronted with a barrage of corporate marketing" (ibid., 439). However, SEPTA has reaped comparatively modest incomes via the naming rights (circa $1,100,000 per annum in 2010–2015 and lately smaller sums). Meanwhile, the "quasi-privatization" process in focus has divested city dwellers' common identifiers and non-market values previously attached to Pattison station.

In the US and subsequently in the rest of the world, the naming rights boom began with major sports venues. As epitomized by the Pattison/AT&T/NRG station case, however, it has since then spread to encompass a range of property subtypes in public ownership and use. Naming rights for metro and other transportation stations, cable cars, schools, training pitches, campus buildings, lecture halls and so forth are increasingly commonly leased for short-turn returns, with varying levels of success. Rose-Redwood et al. (2019a, p. 850) encapsulate the applied policy toolkits at work behind such fiscal gambits:

> The pursuit by urban governments and other institutions for "naming rents" continues to grow and is becoming more professionalized and refined. In many cases, naming rights sales are the result of ad hoc agreements between local governments and individual sponsors. However, an increasing number of cities have adopted formal naming

rights policies, which codify the rules for individual naming rights agreements.

In their study, Rose-Redwood et al. (2019a) focus on two large-scale municipal naming rights programs, issued by the cities of Winnipeg, Canada and Dubai, the UAE. The former was an attempted campaign to sell naming rights for all thinkable types of city assets (from roads, bridges, parkades, swimming pools and community centers to parking meters and "even a police helicopter"); the latter concerned 23 metro stations. Aside from municipalities' growing interest in such en masse campaigns, a more usual privatization procedure has gone through codifying the rules for name leaseholds of publicly owned and funded properties by entrepreneurial local and regional territorial administrations (ibid.; see also Harvey 1989; Lauermann 2016). An institutional upheaving of this type of toponymic commodification occurred, for instance, when the state of Pennsylvania passed an act in 2007 which required transit agencies to find new sources of revenue (Weaver 2019, p. 439). In another type of case concerning a city-owned venue built for the 2002 Commonwealth Games, Manchester City Football Club (by then taken over by Abu Dhabi's ruling Al Nahyan family) "negotiated with the City Council [...] for control of the naming rights in return for an increase in rent payments" in 2010 (Medway et al. 2019, p. 789), which soon led to the renaming of the venue as Etihad Stadium (located within the Etihad campus also eponymous to the Abu Dhabi airline). In the Finnish provincial capital of Oulu, ensuing naming right deals were bolstered up by a council decision in 2003 that main tenants in its premises are allowed to negotiate naming rights contracts on their own. Also in many other cases scanned for this study, either a main tenant, or an entity like a joint venture company or a development corporation, is authorized to carry out negotiations and decisions over a publicly owned facility's name without direct democratic or political control. This type of outsourcing of public governance through legally separate entrepreneurial entities adds an important property relations dimension to what Weaver (2019) called the "quasi-privatization" of publicly owned and communally used spaces (see also Penny 2016).

### 6.3.2. *Sporting venue naming rights as contested cash machines*

In connection with entertainment and spectator sports venues, conditions are especially favorable for lucrative naming rights contracts, often tailored as part of larger sponsorship arrangements (Gillooly and Medway 2018; Vuolteenaho et al. 2019a, p. 764). With the advantage of immense media attention and fan emotions invested on them, venue owners and sporting clubs "are able to advertise place in a way that place marketers can only dream about" (Edensor and Millington 2008,

p. 178). In North America and Germany, in particular, sporting venue naming rights have become a new normal to an extent that not selling a historical venue name can be "a badge of distinction" (Boyd 2000, 2013; Vuolteenaho et al. 2019a).

Hardly nowhere, however, have the realized and projected naming rights deals gone unchallenged (e.g. Woisetschläger et al. 2014; Demuru 2016; Madden 2019). Even in the US and Germany, civic protests like petitioning for the reinstation of the name of (now-demolished) Candlestick Park in San Francisco, campaigning against the corporate name of Invesco Field at Mile High in Denver, or Hamburg's FC St Pauli supporters' stipulation that "the name Millerntor-Stadium cannot be sold, modified or changed", exemplify this countertendency (Merkel 2012, p. 367; Jarvis 2019). At the same time, however, there are fans and locals who accept or tolerate naming rights as a kind of economic "must", or even enthuse over them. Illustrative of these varying civic responses, some posters on the *jatkoaika.com* forum saw the fourth consecutive corporate re-labeling of a council-owned indoor venue in the Finnish city of Turku to *Gatorade Center* (after a sports drink brand of Pepsi Company) as a belated "progress", "air of quality" and "prestige" eventually landed onto the Finnish soil: "What a nice name! Some big world action here! In striking contrast, others could not "in any way see how a name's change into English one in itself represents any kind of 'development'", or contented that the entire practice of changing naming rights "is sheer clownery, a venue should have a name that will not wear thin".

In many parts of Europe, (association) football is "paradoxically known both for its heavyweight commercialism and as a realm of sometimes enthusiastically traditionalist fan cultures" (Vuolteenaho et al. 2019a, p. 765). The latter identity aspect is regularly attached to the game's tradition-rooted landmark shrines and their inherited names. This easily translates into tensions around naming rights sponsorships, whether the stadiums are publicly or privately owned. As a high-end example, an initiative pushed by the Real Madrid football club's Canadian-born head of partnerships to secure a naming rights contract for its iconic Santiago Bernabéu, has been at least for the time being diluted due to investors' fears that the club's core fan base would not embrace "any partnering brand" lending its name to the stadium (Friend 2018).

Occasionally, both popular and political-authoritative forms of opposition are mobilized against a naming rights arrangement. Such was the case in perhaps the most notorious European episode in renaming a time-honored football ground after a corporate power. In this case, St. James' Park (so named since the 1880s) in Newcastle, north-eastern England, was renamed without any consultations twice, first semi-officially to *sportsdirect.com@StJames'Park* (most of the 2009–2010 season) and then officially to *Sports Direct Arena* (11/2011–10/2012). In charge of the name changes was nominally the Newcastle United FC (NUFC) and its holding company (authorized to re-naming as the council-owned venue's main tenant).

In practice, the terms were set by the club owner Mike Ashley, also the proprietor of the sportswear and sports equipment retailer Sports Direct. Technically, neither of the new corporatized names were the result of a dyadic name sponsorship deal, but rather of intra-company decisions to "showcase" the venue name's marketing potential to sponsor candidates. NUFC supporters' immediately counteracted, including furious online storming and direct action on the streets (Medway et al. 2019).

A preponderance of discussants on forums like www.nufcblog.org believed that London-based Ashley was just marketing his own Sports Direct empire, speculatively maneuvering at the cost of an iconic name engrained into people's Geordie identities. In both re-naming cases, Newcastle's council was on the protestors' side. It appealed to the media to use the name St. James' Park only, and refused to change street and transit station signage in the stadium's vicinity. Twelve Labour MPs signed a House of Commons motion to cut short "corporate vandalism" and "total disrespect for the history and culture of our region"[5]. These multifront protests were to no avail at first. Showing the overbearing power of business contracts in our neoliberal times, things only changed in October 2012 when the payday loan company Wonga entered the scene as the club's new main sponsor (covering shirt sponsorship and stadium naming rights). In its own publicity maneuver, Wonga – a widely criticized "legal loan shark" (Spellman 2012) – announced its "gift" to NUFC fans: they were handed "back their home of football by reinstating the stadium's rightful St. James' Park name" (Newcastle United 2012). The protestors had won a bitter battle against the corporatization of the name of their beloved sanctum, at least temporarily.

### 6.3.3. *Glimpses at rental variations in the naming rights marketplace*

Large-scale comparative studies on the spread of naming rights-style toponymic commodification remain scarce, but salient instances of it are found in all continents (see a comparative study in six European countries: Vuolteenaho et al. 2019a). Nonetheless, fleeting observations on multi-scalar naming rights-related power geometries and inequalities have been made (Rose-Redwood et al. 2019b, p. 758). Following metropolises, smaller localities in many national contexts have jumped the naming rights bandwagon (Boyd 2000; Vuolteenaho et al. 2019a). European

5. https://www.bbc.com/news/uk-england-tyne-17123563.

evidence suggests that there is a strong tendency of name sponsorships to last considerably shorter in peripheral or semi-peripheral geographical settings with more impecunious local economies, smaller edifices, lowly ranked sports clubs and less media attention. In smaller Finnish and Scottish towns with stagnating or declining economies, for instance, the contracts typically last only for a couple of years or even less (ibid., 776). In contrast, naming rights contracts valid for between 10 and 30 years are characteristic to properties located in bigger cities.

When it comes to the rental value variations, cleavages in the naming rights market are gaping even wider, showing that "rent is determined relationally over all spheres of production in all locations, with future expectations also incorporated into the calculation" (Harvey 1973, p. 186). The highest sums paid for naming rights are exorbitant. Abu Dhabi's recent pursuits to worldwide recognition through spatial brand name loans such as Sorbonne University Abu Dhabi, Guggenheim Abu Dhabi and Louvre Abu Dhabi, open here illustrative comparison points. In a franchise-type of deal between the city of Abu Dhabi and the French government, the former paid the latter USD 525 million for a 30-year right (17.5 million per a year) of its Abu Dhabi Louvre art museum to be associated with the iconic Louvre superbrand of high art and cultural tourism (in this type of franchises cash flows from the real property owners to trademark proprietors).

Especially in the hubs of the financialized real estate capitalism (e.g. Slater 2017), regular types of naming rights contracts for indoor arenas, for instance, also boast eight-figure annual USD sums. In a 2007 deal welcomed by the NYC deputy mayor as "a remarkable vote of confidence", Brooklyn's Barclays Center has yielded 10 USD million yearly income from a London-headquartered banking institution to a partnership running this world-class event facility (in this case, the original deal at USD 400 million for 20 years was even scaled down to a half during the late-2000s slump). This one-time record sum astoundingly trumps more down-to-earth naming rent values both in the venue's vicinity, and especially in remoter locations bypassed big global capital flows. In a coeval name sponsorship for the adjacent Atlantic Avenue–Barclays Center, New York City's Metropolitan Transport Authority was able to negotiate a receipt of "just two hundred thousand dollars per annum" (Rose-Redwood and Alderman 2011, p. 3; Light and Young 2015, pp. 441–442; Weaver 2019, p. 438). Hence, the historic subway station complex yielded a bare two per cent of the annual premium of the adjacent spectacular entertainment and sports venue, a stark naming rent difference hinting at divergence between "public austerity and private luxury" (Madden 2019, 891) in the local real estate landscape. Out of Winnipeg and Dubai's (Rose-Redwood et al. 2019a) aforementioned campaigns, only the latter got close to the upper crust with its first 13 metro stations name leaseholds valued at USD 489 million (with a minimum duration set at 10 years, an average of USD 3.8 million/station/year at most). Winnipeg's campaign backfired: naming rights for miscellaneous city assets generated pittances (an aggregate of CAD 396,200 in 2011 and CAD 138,900 in 2015) compared to municipal policymakers' wishful expectations at CAD 1.5 million per annum (ibid., 859).

At the millennium's turn, prices that sponsors paid for a major league stadium or arena name in the US were in the region of USD 2 million per annum (McCarthy and Irwin 2000). In my native Finland, legislation allows the naming rights price tags to be kept as trade secrets. In the upper end of the spectrum, in one reported case from 1994, a Finnish beverage manufacturer purchased a right to call the country's then biggest indoor venue *Elysée Arena* (a preceding sponsored name of the aforementioned Gatorade Center) for USD 120,000 per year. In a more recent two-year deal, the City of Espoo in the capital region paid USD 280,000 per annum for the Espoo Metro Arena's name in an effort to keep the venue's private operator afloat. Even today, severely humbler five-figure annual amounts are being paid for re-branding indoor arenas toponymically in the country's stagnating provincial municipalities. With its dozens of cut-rate and a couple of somewhat more lucrative contracts, Finland clearly belongs to the planet's "naming-rights-peripheries". In the naming rights marketplace, egregious power disparities between the globalized economy's hubs and fringes exist, signaled both by durations of contracts and sums involved.

## 6.4. Discussion: the property–name nexus as a commodification frontier

Through instances of thematic branding and spatial naming rights, we have seen that the processes of toponymic commodification do not occur in a vacuum, but as part and parcel of profit-driven and power-infused productions of space. To paraphrase Henri Lefebvre (1991, p. 360), a "lot" – a property conceived as a piece of land and the real estate on it – is the fundamental unit in the functioning of the space of capitalism, and particularly its real estate markets. In pre- and proto-capitalist societies, by contrast, "markets existed but not in land" (Obeng-Odoom 2016), and the parceling of space into exclusive and alienable private possessions was partial and exceptional as landownership was overlapping (in the classical case of pre-enclosure rural England, "the commoners had rights to the same piece of land as the king, the lords and tenants"; Haila 2016, p. 30). Only in the wake of the capitalist modernization, starting from the West, space/land became treated as a commodity, a possession of a single owner and an object to be bought and sold (e.g. Polanyi 1944–2001; Macpherson 1978; Ward 1998; Haila 2016).

Crucially, in jurisprudential terms, the word "exclusive" denotes here the owner's principally "full" rights to legitimately act (to sell, develop, subdivide, fence, make contracts, *name*, draw rent, speculate and so on) on a place property – irrespective of whether the titled proprietor is an individual, a corporation or a governmental body (Blomley 2004). This commodified logic is the theoretical point of departure of the concept of property–name nexus proposed by this chapter. Bluntly put, what this chapter's property relations-attuned investigations suggest is that the exclusive single-owner property model is the launchpad of toponymic commodification, both past and present (see also Plumwood 2005).

In each of this chapter's cases, a private or quasi-private actor or partnership – with more-or-less exclusive rights to name and act on its property, and to a varying degree in collaboration with authorities – imposed its or its sponsor's brand visions onto the namescape. Out of the discussed examples, the purest toponymic manifestation of exclusive property rights was the pre-neoliberal idyllic namescape planted within Sun City (see also Aalbers 2017). After it had acquired a former cotton farm, the way in which the DEVCO "manipulated the place for its exchange values" (Logan and Molotch 2007), and put its business vision into "toponymic action" was the company's prerogative only (in a very dissimilar South-East Asian political-economic context, the Vinhomes Central Park also came close to this plain version of property–name nexus). No less notably, the instrumental logic of naming rights can also be conceptualized as a specifically neoliberal outgrowth and application of exclusive property rights[6]. In this contractualized nomenclature, a place name is reconfigured as an ingredient of the owner's rights to a place property, and literally commodified as a transferable right to name – as something that the owner (or any legally separate entity authorized by it) can put up for sale (see also Rose-Redwood et al. 2019a, pp. 850–851). The whole quid pro quo logic of naming rights agreements between the seller (garnering additional rent from one's property) and the purchaser (gaining publicity benefits) would be unthinkable – as has been the case in most parts of the world until the last three decades – without the hegemony of the single-owner property doctrine in societies of "advanced" neoliberal capitalism (see also Madden 2019, p. 889).

6. While more research on the ideological, intellectual and institutional foundations behind this corporatized toponymic language is needed (see, for example, Rose-Redwood 2019a, p. 848), the now globalized naming rights discourse is in line with the tenets of the influential Law and Economics subfield (also known as the Property Rights School) of the Chicago School of Economics (e.g. Coase 1991; see, critically, Beatley 1989; Haila 2005). Demsetz (1967) argued that property rights have an important function "in guiding incentives to achieve greater internalization of externalities" (Haila 2016, pp. 53–55), translatable as commodification also in toponymic terms.

While influential economists and ideologists of capitalism from John Locke to Milton Friedman and Ronald Coase defended the private ownership model as a foundation of a "free society", political economists insist that actual ownership forms are inescapably embedded in the wider constellations of power, socio-spatial relations and citizens' rights (Harvey 1973; Blomley 2004; Haila 2016). Throughout capitalist modernization, the hegemony of single-owner property rights has been predicated on complex interdependencies between the market and the state (e.g. Polanyi 1944–2001; Ward 1998)[7]. The state has not been an inactive bystander of the triumph of the private ownership model, but its most influential guarantor. The state's final say over property relations (including the property–name nexus) is nakedly exposed in the processes of eminent domain (expropriations of private property), and also during megaevents when a host state's contract-based responsibilities may temporarily override existing venue naming rights contracts. Central and local governments have had, and continue to have, vested political-economic, strategic and fiscal interests in consolidating the private ownership and profit-seeking, and also legal, technical and policy tools at their disposal to pursue such ends (Elden 2013; Sevilla-Buitrago 2015).

In the neoliberal-era cases discussed in this chapter, state action took multiple forms, which in some cases fed into rather complicated processes of toponymic commodification. The St. James' Park's attempted but short-lived renaming showed that not all policymakers invariably welcome the commodified enclosures of toponymy. Local counselors and authorities in Newcastle joined the civic resistance against the historic property's threatened "publicness" vis-à-vis the tenant's speculative naming initiatives. However, this oppositional stance-taking to a commodified property–name nexus by policymakers was occasion-specific and otherwise absent from the discussed examples. More consistently, the scanned local states (including Newcastle) had outsourced or quasi-privatized the governance of publicly owned venues, thereby delimiting democratic oversight over the selling of naming rights (see also Penny 2016; Madden 2019, p. 889). As a proliferating type of fiscal technique, Winnipeg and Dubai's decisions to put public infrastructure names up for en masse sale epitomized entrepreneurial municipalities' fiscal prioritizations (Rose-Redwood et al. 2019a). Ho Chi Minh City's investment-hungry

7. Macpherson (1978) goes as far as to suggest that state property itself is "a form of corporate private property" (cited in Haila 2016, 216). In analytical terms, to apply this theoretical position literally and so as to lump together the (usually state-dictated) street toponymy and the (state-sanctioned) commercial toponymy is hardly commendable. However, it is conceptually expedient to separate both the "statist" street toponymy and "private" commodified toponymy from self-organized toponymic "commons" that evade both governmental control and the logic of exchange value (e.g. Hardt and Negri 2009; Huron 2015). I leave the toponymic elaborations of this conceptual triplet for future research.

land selling policy paved the way for the "westernized" neotoponymy of Vinhomes Central Park to emerge. In Aurinkolahti's case – where again no naming rights per se but parcels of public land were traded (Haila 2005) – the city's maneuver to suspend its planning monopoly, and instead authorize a local developer to captain the area's development, yielded a conspicuously market-oriented thematic namescape (Vuolteenaho and Ainiala 2009, p. 247). Both private and public property owners in localities of varying economic clout had joined the neoliberal naming rights bandwagon, which was manifested as egregious naming rent differences between the hubs and (semi)peripheries of the transnational naming rights marketplace. In all of these cases, city governments acted like much like exchange value-driven private property owners, with far-reaching implications onto the commodification of local namescapes.

## 6.5. References

Aalbers, M.B. (2017). The pre-histories of neoliberal urbanism in the United States. In *Debating the Neoliberal City*, Morel Journel, C. and Pinson, G. (eds). Ashgate, Aldershot.

Alderman, D. and Ingwood, J. (2018). Street naming and the politics of belonging: Spatial injustices in the toponymic commemoration of Martin Luther King, Jr. In *The Political Life of Urban Streetscapes: Naming, Politics, and Place*, Rose-Redwood, R., Alderman, D., Azaryahu, M. (eds). Routledge, Abingdon.

Beatley, T. (1994). *Ethical Land Use*. Johns Hopkins University Press, Baltimore.

Beauregard, R.A. (2006). *When America Became Suburban*. University of Minnesota Press, Minneapolis.

Berg, L.D. (2011). Banal naming, neoliberalism, and landscapes of dispossession. *ACME: An International E-Journal for Critical Geographies*, 10(1), 13–22.

Blomley, N. (2004). *Unsettling the City: Urban Land and the Politics of Property*. Routledge, London.

Bourdieu, P. (1986). The forms of capital. In *Handbook of Theory and Research for the Sociology of Education*, Richardson, J. (ed.). Greenwood, New York.

von Borries, F. (2004). *Who's Afraid of Niketown? Nike-urbanism, Branding and the City of Tomorrow*. Episode Publishers, Rotterdam.

Boyd, J. (2000). Selling home: Corporate stadium names and the destruction of commemoration. *Journal of Applied Communication Research*, 28(4), 330–346.

Boyd. J. (2013). Cheering for teams that don't sell stadium names. Lambeau Field, Fenway Park – Honoring tradition is a badge of distinction. *The Wall Street Journal*, 18 August [Online]. Available at: https://www.wsj.com/articles/cheering-for-teams-that-dont-sell-stadium-names-1376862026.

Brenner, N. and Theodore, N. (2002). *Spaces of Neoliberalism*. Blackwell, Malden.

Burton, T. (2008). *Naming Rights: Legacy Gifts and Corporate Money*. Wiley, Hoboken.

Butterworth, M.L. and Moskal, S.D. (2009). American football, flags, and "fun": The bell helicopter armed forces bowl and the rhetorical production of militarism. *Communication, Culture & Critique* 2, 411–433.

Caiazzo, L., Coates, R., Azaryahu, M. (2020). *Naming, Identity and Tourism*. Cambridge Scholars Publishing, Newcastle.

Chen, L.N.H. (2017). Names of Chinese skyscrapers. *Names*, 65(1), 36–44.

Christophers, B. (2010). On voodoo economics: Theorising relations of property, value and contemporary capitalism. *Transactions of the Institute of British Geographers*, 35, 94–108.

Christophers, B. (2018). *The New Enclosure: The Appropriation of Public Land in Neoliberal Britain*. Verso, London.

Coase, R.H. (1991). The institutional structure of production. The 1991 Alfred Nobel Memorial Prize lecture in economic science. R.H. Coase Essays on Economics and Economists. The University of Chicago Press, Chicago.

Creţan, R. (2019). Who owns the name? Fandom, social inequalities and the contested renaming of a football club in Timişoara, Romania. *Urban Geography*, 40(6), 805–825.

Demsetz, H. (1967). Towards a theory of property rights. *American Economic Review*, 57, 347–359.

Demuru, P. (2016). Arenas, naming rights e apelidos: Estratégias discursivas e modelos de cidade na comunicação do Allianz Parque e da Arena Corinthians. *Intexto*, 37, 292–312.

Doling, T. (2019). *Exploring Saigon-Chợ Lớn: Vanishing Heritage of Hồ Chí Minh City*. Thế Giới Publishers, Hà Nội.

Edensor, T. and Millington, S. (2008). This is our city: Branding football and local embeddedness. *Global Networks*, 8(2), 172–193.

Elden, S. (2013). *The Birth of Territory*. University of Chicago Press, Chicago.

Findlay, J.M. (1992). *Magic Lands: Western Cityscapes and American Culture After 1940*. University of California Press, Berkeley.

Fortunato, J.A. (2013). *Sports Sponsorship: Principles and Practices*. McFarland & Company, Jefferson.

Fraser, N. (2014). Can society be commodities all the way down? Post-Polanyian reflections on capitalist crisis? *Economy and Society*, 43(4), 541–558.

Friend, N. (2018). Real Madrid struggle for bernabeu naming rights deal. *SportsPro*, 10 October [Online]. Available at: https://www.sportspromedia.com/news/real-madrid-bernabeu-naming-rights-deal.

Gillooly, L. and Medway, D. (2018). Sponsorships, stadia and naming rights. In *Routledge Handbook of Football Business and Management*, Chadwick, S., Parnell, D., Widdop, P., Anagnostopoulos, C. (eds). Routledge, London.

Giraut, F. and Houssay-Holzschuch, M. (2016). Place naming as dispositif: Toward a theoretical framework. *Geopolitics*, 21(1), 1–21.

Gottdiener, M. (2001). *The Theming of America*. Westview Press, Oxford.

Haila, A. (1999). City building in the East and West: United States, Europe, Hong Kong and Singapore compared, *Cities*, 16(4), 259–267.

Haila, A. (2005). Kahvinpaahtaja kaupunkisuunnittelijana. In *Suunnittelun kumppanuudet. Tapaus Vuosaari*, Bäcklund, P. and Schulman, H. (eds). Helsingin kaupungin tietokeskus, Helsinki.

Haila, A. (2016). *Urban Land Rent: Singapore as a Property State*. Wiley-Blackwell, Oxford.

Hardt, M. and Negri, A. (2009). *Commonwealth*. Harvard University Press, Cambridge.

Harms, E. (2012). Beauty as control in the new Saigon: Eviction, new urban zones, and atomized dissent in a Southeast Asian city. *American Ethnologist*, 39(4), 735–750.

Harms, E. (2019). Megalopolitan megalomania: Ho Chi Minh City, Vietnam's Southeastern region and the speculative growth machine. *International Planning Studies*, 24(1), 53–67.

Harvey, D. (1973). *Social Justice and the City*. Johns Hopkins University Press, Baltimore.

Harvey, D. (1989). From managerialism to entrepreneurialism: The transformation in urban governance in late capitalism. *Geografiska Annaler B*, 71, 3–17.

Harvey, D. (2012). *Rebel Cities: From the Right to the City to the Urban Revolution*. Verso, London.

Hopkins, J.S.P. (1990). West Edmonton Mall: Landscape of myths and elsewhereness. *The Canadian Geographer*, 35(3), 2–17.

Huron, A. (2015). Working with strangers in saturated space: Reclaiming and maintaining the urban commons. *Antipode*, 47(4), 963–979.

Jarvis, C.M. (2021). Navigating the public-place/private-market tension: Toponymic and material reading of Broncos Stadium. *Sport in Society*, 24(3), 473–489.

Johansson, O. and Cornebise, M. (2010). Place branding goes to the neighbourhood: The case of pseudo–Swedish Andersonville. *Geografiska Annaler: Series B, Human Geography*, 92(3), 187–204.

Karimi, A. (2016). Street fights: The commodification of place names in post-Taliban Kabul City. *Annals of the American Association of Geographers*, 106(3), 738–753.

Kearns, R.A. and Lewis, N. (2019). City renaming as brand promotion: Exploring neoliberal projects and community resistance in New Zealand. *Urban Geography*, 40(6), 870–887.

Klingmann, A. (2007). *Brandscapes*. MIT Press, Cambridge, MA.

Lauermann, J. (2016). Municipal statecraft: Revisiting the geographies of the entrepreneurial city. *Progress in Human Geography*, 42(2), 205–224.

Leary, M.E. (2009). The production of space through a shrine and vendetta in Manchester: Lefebvre's spatial triad and the regeneration of a place renamed castlefield. *Planning Theory & Practice*, 10(2), 189–212.

Lefebvre, H. (1991–1974). *The Production of Space*. Blackwell, Oxford.

Light, D. (2014). Tourism and toponymy: Commodifying and consuming place names. *Tourism Geographies*, 16(1), 141–156.

Light, D. and Young, C. (2015). Toponymy as commodity: Exploring the economic dimensions of urban place names. *International Journal of Urban and Regional Research*, 39(3), 435–450.

Logan, J.R. and Molotch, H.L. (2007). *Urban Fortunes: The Political Economy of Place*, 20th Anniversary Edition. University of California Press, Berkeley.

Macpherson, C.B. (1978). *Property: Mainstream and Critical Positions*. Basil Blackwell, Oxford.

Madden, D.J. (2014). Neighborhood as spatial project: Making the urban order on the downtown Brooklyn waterfront. *International Journal of Urban and Regional Research*, 38(2), 471–497.

Madden, D.J. (2018). Pushed off the map: Toponymy and the politics of place in New York City. *Urban Studies*, 55(8), 1599–1614.

Madden, D.J. (2019). The names of urban dispossession: A concluding commentary. *Urban Geography*, 40(6), 888–892.

McCann, E. (2013). Policy boosterism, policy mobilities, and the extrospective city. *Urban Geography*, 34(1), 5–29. DOI: https://doi.org.10.1080/02723638.2013.778627.

McCarthy, L. and Irwin, R. (2000). An examination of the rationale and motives for corporate purchase of stadia and arena naming rights. *The Cyber-Journal of Sport Marketing*, 4(3), 1–9.

McElroy, E. (2019). Data, dispossession, and Facebook: Technoimperialism and toponymy in gentrifying San Francisco. *Urban Geography*, 40(6), 826–845.

Medway, D. and Warnaby, G. (2014). What's in a name? Place branding and toponymic commodification. *Environment and Planning A*, 46, 153–167.

Medway, D., Warnaby, G., Gillooly, L., Millington, S. (2019). Scalar tensions in urban toponymic inscription: The corporate (re)naming of football stadia. *Urban Geography*, 40(6), 784–804.

Merkel, U. (2012). Football fans and clubs in Germany, conflicts, crises and compromises. *Soccer & Society*, 13(3), 359–376.

Miles, S. (2010). *Spaces for Consumption*. Sage, London.

Newcastle United (2012). Toon fans celebrate as St. James' reborn. 17 October [Online]. Available at: https://www.nufc.co.uk/news/archive/toon-fans-celebrate-as-st-james-reborn/.

Norris, D.A. (1999). Unreal estate: Words, names and allusions in suburban home advertising. *Names*, 47(4), 365–380.

Obeng-Odoom, F. (2016). Marketising the commons in Africa: The case of Ghana. *Review of Social Economy*, 74(4), 390–419.

Pacione, M. (2012). The retirement village as a residential environment for the third age – The example of Firhall, Scotland. *Scottish Geographical Journal*, 128, 2.

Peck, J. (2015). Cities beyond compare? *Regional Studies*, 49(1), 160–182.

Penny, J. (2016). The (post)politics of fiscal retrenchment. In *Urban Austerity: Impacts of the Global Financial Crisis on Cities in Europe*, Schönig, B. and Schipper, S. (eds). Theater in Zeit, Berlin.

Plumwood, V. (2005). Belonging, naming and decolonisation. In *Habitus: A Sense of Place*, Hillier, J. and Rooksby, E. (eds). Routledge, London.

Polanyi, K. (1944–2001). *The Great Transformation: The Political and Economic Origins of Our Time*. Beacon Press, Boston.

Raento, P. and Douglass, W.A. (2001). The naming of gaming. *Names*, 48(1), 1–31.

Rose, C. (1986). The comedy of the commons: Custom, commerce, and inherently public property. *University of Chicago Law Review*, 53(3), 711–781.

Rose, C. (1994). *Property and Persuasion. Essays on the History, Theory, and Rhetoric of Ownership*. Westview Press, Boulder.

Rose-Redwood, R. and Alderman, D. (2011). Critical interventions in political toponymy. *ACME: An International E-Journal for Critical Geographies*, 10(1), 1–6.

Rose-Redwood, R., Alderman, D., Azaryahu, M. (2010). Geographies of toponymic inscription: New directions in critical place-name studies. *Progress in Human Geography*, 34(4), 453–470.

Rose-Redwood, R., Sotoudehnia, M., Tretter, E. (2019a). Turn your brand into a destination: Toponymic commodification and the branding of place in Dubai and Winnipeg. *Urban Geography*, 40(6), 846–869.

Rose-Redwood, R., Vuolteenaho, J., Young, C., Light, D. (2019b). Naming rights, place branding, and the tumultuous cultural landscapes of neoliberal urbanism. *Urban Geography*, 40(6), 747–761.

Sevilla-Buitrago, A. (2015). Capitalist formations of enclosure: Space and the extinction of the commons. *Antipode*, 47(4), 999–1020.

Slater, T. (2017). Planetary rent gaps. *Antipode*, 49, 114–137.

Soja, E.W. (1989). *Postmodern Geographies*. Verso, London.

Sorkin, M. (1992). *Variations on a Theme Park: The New American City and the End of Public Space.* Hill and Wang, New York City.

Sotoudehnia, M. (2018). Toponymic checksum or flotsam? Recalculating Dubai's grid with Makani, "the smartest map in the world". In *The Political Life of Urban Streetscapes: Naming, Politics, and Place*, Rose-Redwood, R., Alderman, D., Azaryahu, M. (eds). Routledge, Abingdon.

Spellman, D. (2012). Newcastle stadium name returns to St James' Park after four-year sponsorship deal with wonga.com is struck. *Independent*, 9 October [Online]. Available at: https://www.independent.co.uk/sport/football/premier-league/newcastle-stadium-name-returns-to-st-james-park-after-four-year-sponsorship-deal-with-wongacom-is-8203956.html.

Spocter, M. (2018). A toponymic investigation of South African gated communities. *South African Geographical Journal*, 100(3), 326–348.

Tegelman, T. (2006). Kaupunkitaiteilijan arjen ja nimettyjen paikkojen kohtaamiset Helsingin Aurinkolahdessa ja Kalliossa. Unpublished MSc Thesis, University of Helsinki, Helsinki.

Trinch, S. and Snajdr, E. (2017). What the signs say: Gentrification and the disappearance of capitalism without distinction in Brooklyn. *Journal of Sociolinguistics*, 21(1), 64–89.

Tullock, G. (2005). *The Rent-Seeking Society.* Liberty Fund, Indianapolis.

Vuolteenaho, J. and Ainiala, T. (2009). Planning and revamping urban toponymy: Ideological alterations in the linguistic landscaping of Vuosaari suburb, eastern Helsinki. In *Critical Toponymies: The Contested Politics of Place Naming*, Berg, L.D. and Vuolteenaho, J. (eds). Ashgate, Farnham.

Vuolteenaho, J. and Kolamo, S. (2012). Textually produced landscape spectacles? A Debordian reading of Finnish namescapes and English soccerscapes. *COLLeGIUM: Studies across Disciplines in the Humanities and Social Sciences*, 13, 132–158.

Vuolteenaho, J., Wolny, M., Puzey, G. (2019a). This venue is brought to you by...: The diffusion of sports and entertainment facility name sponsorship in urban Europe. *Urban Geography*, 40(6), 762–783.

Vuolteenaho, J., Lappalainen, H., Ainiala, T. (2019b). A slightly slummier area? Negotiations of place-bound identities through social spatializations and unofficial toponyms. *Journal of Sociolinguistics*, 23(2), 140–162.

Ward, S.V. (1998). *Selling Places: The Marketing and Promotion of Towns and Cities 1850–2000.* E&FN Spon, London.

Weaver, T.P.R. (2019). The new enclosures: London, New York City, Philadelphia, and the transformation of public space. *New Political Science*, 41(3), 423–442.

Wideman, T.J. and Masuda, J.R. (2018). Assembling "Japantown"? A critical toponymy of urban dispossession in Vancouver, Canada. *Urban Geography*, 39(4), 493–518.

Woisetschläger, D.M., Haselhoff, V.J., Backhaus, C. (2014). Fans' resistance to naming right sponsorships: Why stadium names remain the same for fans. *European Journal of Marketing*, 48(7/8), 1487–1510.

Wood, D. (2002). Selling the suburbs: Nature, landscape, adverts, community. *Transformations* [Online]. Available at: http://transformations.cqu.edu.au/journal/issue5/articles/text.htm.

Yakhlef, A. (2004). Global brands as embodied "generic spaces". *Space & Culture*, 7(2), 237–248.

Yurchak, A. (2000). Privatize your name: Symbolic work in a post-Soviet linguistic market. *Journal of Sociolinguistics*, 4(3), 406–434.

Zhao, B., Huang, X., Sui, D.Z. (2019). Place spoofing: A case study of the Xenophilic copycat community in Beijing, China. *The Professional Geographer*, 71(2), 265–277.

# 7

# The Toponymy of Tourism and Leisure: General Framework and Lessons from France

**Christophe GAUCHON**
*Edytem, Université Savoie-Mont-Blanc, Chambéry, France*

> "Isn't it natural to call our baths: the Bains du Mont-Oriol, and to attach to this resort, which will become one of the most important in the whole world, the name of the first owner. Render unto Caesar that which is Caesar's. And note, gentlemen, that this term is excellent. We will say Mont-Oriol, as we say Mont-Dore. It lingers in the eye and in the ear, we see it well, we hear it well, it stays with us: Mont Oriol! 'Mont Oriol!' – The baths of Mont-Oriol…" Guy de Maupassant, *Mont-Oriol*, 1886, 1st part, Chapter 8.

## 7.1. Introduction

Thus, in France at the end of the 19th century, Guy de Maupassant highlighted the links between the touristic positioning of places and toponymic production. In his novel *Mont-Oriol*, a financier named Andermatt, promoter of the new resort in the Auvergne, affirms loud and clear: "I want to launch a spa town!" and, leaving nothing to chance, he integrates the toponymic dimension into his launch campaign, which allows him to flatter the Oriols, father and son, owners of the land from where

*The Politics of Place Naming*,
coordinated by Frédéric GIRAUT and Myriam HOUSSAY-HOLZSCHUCH. © ISTE Ltd 2022.

the spring flows. This new name ensures its success by conforming to a toponymic model then in vogue, hence the comparison with Mont-Dore. Following the same process, in the 1970s, the names of *Val Thorens*, *Valmorel* and *Valfrejus* were created in the Alps, echoing the precedent of Val d'Isère.

Almost a century after Maupassant, Samivel (1967) echoes him in *The Madman of Edenberg* where the transformation of the small village of *Saint-Beat* into a winter sports resort is accompanied by its change of name, first to *Saint-Beat-Edenberg*, then to *Edenberg-Saint-Beat*, and finally to *Edenberg*. However, this is exactly what happened, for example, in Loire-Atlantique in another classic French case of tourism development: in the 1920s, the town of *Escoublac* thus became *Escoublac-La Baule*, then *La Baule-Escoublac* in the 1960s, and the tourist resort itself is more commonly referred to by the name of *La Baule*.

The underlying idea is that the invention of a new place entails the production of a new name. Geographers of tourism have long identified this link: Remy Knafou et al. accordingly evoked this "invention of the toponymy of hitherto undistinguished places", but without giving a specific example and without expanding on this aspect (1997, p. 196). Before them, B. Debarbieux and H. Gumuchian had taken an interest in the names given to ski slopes and places in the high alpine mountains, shedding light on the socio-cultural mechanisms of their production and the lexical fields involved, insisting above all on the images and the representations that interact with these new names (1987a, 1987b).

In 2014, Duncan Light brought the emerging concerns of critical toponymy into English-speaking academic literature, including the commercialization of place names and the issue of tourist place names. He thus considered the processes of assimilation of a name to a tourist attraction and of a tourist practice to a famous name. In the United States, Route 66 has thus become emblematic of a certain form of automobile tourism, whose development it has supported, and which continues today by virtue of this name synonymous with freedom (Light 2014). Hollywood also appears as an exemplary, if not paradigmatic, case of a local place name that has gained global notoriety as an iconic location for the entertainment industry and related tourism (Braudy 2011).

As for neo-toponymy, understood as the production of place names, in other words toponymic invention and power, we can say that tourist neo-toponymy is part of the more general movement of subversion of places by tourism (Équipe MIT 2002, p. 219), which profoundly modifies their appearance, diverting them from their primary functions and redirecting them towards others. The toponymy indicates that this subversion not only takes place in the field of practices and developments, but also affects the very meaning of places and the values attached to them, as well as the actors behind these changes. Thus Mona Ozouf, in her memories of childhood on the Brittany coast, evokes "the dark beacon rock of Skeiz – the summer visitors have renamed it the Lion's Head" (Ozouf 2009, p. 78). If the development of tourism does not necessarily lead to changes in pre-existing toponyms, it is almost always accompanied by the production of new toponyms which will contribute to shaping a new symbolic landscape, or even a landscape of identity (Caiazzo et al. 2020). Thus, Light (2014) recalls that MacCannell (1989) identified naming as part of the process of defining tourist attractions: "MacCannell's work is important for its recognition that names can be of interest to tourists and that such names have a broader role in signifying a tourist site/sight." (Light 2014, p. 142)

These new names attributed to tourist sites are not deployed in a vacuum; they take place in a context of poorly framed and often unjustified toponymic production, and far beyond the field of tourism alone. In other words, these new names are all tools of territorial marketing which also knows how to make use of pre-existing place names by inventing new variations and playing on their notoriety. A whole panoply of linguistic processes is thus implemented, including translation, metaphors, puns, inflections and nominalization.

These variations also illustrate the notion of toponymic attractor: not all place names are equally valuable in the symbolic marketplace, and some, by the prestige attributed to them, take precedence over others. To take these different aspects into account, it will therefore be necessary to first observe the mysteries of toponymic production itself before reflecting on changes in the status and uses of place names. Thus, we may consider the naming of tourist places as an element of the territorial recompositions in progress.

The case of France appears to be particularly rich in terms of illustrations, and sometimes of models, for the different tendencies at work in tourism neo-toponymy. In fact, in certain coastal or mountain regions, place naming has undergone a significant evolution over more than 150 years under the impetus of tourism. The stakes of representation, the concern to identify the territory as a tourist destination, the instrumentalization of place names have resulted in an abundant production of place names. Tourism is not the only cause, and it converges with other interests of local and regional authorities who are involved in toponymic production for tourism purposes.

## 7.2. The new names of tourist places

The name of the *Côte d'Azur*, launched in 1888 by Stephen Liégeard, designated a specific entity which had no prior history as such. It gave new visibility to a then-emerging tourist region, and conferred a certain unity on the coast of the Alpes-Maritimes which, 30 years earlier, had been divided between the French Var coast and the county of Nice belonging to the Kingdom of Piedmont-Sardinia. Toponymic innovation therefore reflects both an economic dynamic and a geopolitical change. Since then, the *Côte d'Azur* has become permanently established in the geographical nomenclature, appropriated by tourists, locals and institutions alike, even in the name of the region *Provence-Alpes-Côte d'Azur*; even though it sometimes faces competition from the more international name of the *Riviera*, or the *French Riviera*, but without this duplicate naming diminishing the glory of the *Côte d'Azur*.

The *Côte d'Azur* therefore remains the archetype of the place name engendered and enshrined by the development of tourism. In 1962, the private consortium preparing to launch the tourist development of Gallura, in the north-east of Sardinia, was created under the name of *Costa Smeralda,* which name was from then on applied to this stretch of the Sardinian coast and became anchored in the tourist geography. Therefore, following the example of the *Côte d'Azur*, a crowd has gathered of place names linked to their tourist functions. And their large number is coupled with a great diversity, of which an initial typology can give an idea:

– Borrowing from pre-existing local place names but transferring the toponym to the tourist place, thus modifying its status and its meaning. Winter sports resorts offer some examples, such as *Norma*, in Maurienne (France), named after a peak (2918 m) which dominates the place chosen for the establishment of the resort.

– Creation ex nihilo of a new place name, without spatial anchoring but based on signifiers or on sounds perceived by the promoters as adding value. Amusement parks like *Mirapolis*, or large ski areas such as *les Sybelles* or *Paradiski*, illustrate these compositions which can be entrusted to firms specializing in *naming* or *branding*.

– Remobilization of pre-existing place names, or sometimes adjectives associated with place names, by metaphorical transposition: since 2016, the Aix-les-Bains tourist office has orchestrated a promotional campaign around Lac du Bourget, renamed for the occasion the *Riviera of the Alps*. Metaphor is a figure that offers a certain flexibility, sometimes playing on the specific place name itself (e.g. *Venise provençale* (the Venice of Provence) to designate Martigues), sometimes on the adjective (e.g. *Drôme provençale* (the Drome of Provence) to designate the southern part of the department), but many of these transpositions struggle to become established over time and barely survive their launch.

– Introduction of franchise-like names which will be used in several places, such as the *Centerparcs* (six locations in the northern half of France) or the *Walibi* leisure parks (four parks in Western Europe), which are self-sufficient at the level of their region of establishment, without any need to specify their location further. Yet the Disney group, despite its universal notoriety, generates specific toponyms for its various parks, with, for example, its Disneyland Paris, initially called EuroDisney Resort, which already differentiated it from the eponymous Disneyland Resort in California or the Walt Disney World Resort in Orlando, which is the reference park for the group (Kaikati and Kaikati 2003).

These neo-toponyms linked to tourism and leisure may be observed at all levels and may be applied to tourist regions (*Costa Smeralda* in Sardinia), to resorts (*Port Camargue*), sectors within resorts, tourism or leisure facilities (names of hotels, beaches, ski lifts) or sites open to visitors. Similarly, they are likely to involve toponyms of all statuses: official toponymy, informal toponymy as well as commercial toponymy, and this great ubiquity explains their universal expansion. Following Boyer and Viallon (1994), Dann (1996) and Light (2014), we can thus underline the role of names of places, sites, destinations or attractions in the creation or recreation of a product image: the place name communicates the appropriate message in a given context to attract the consumer.

Commercial and advertising logics are behind the changes, sometimes with little reason, in these tourist toponyms. Thus, in recent years, the names of Courchevel 1550, 1650 and 1850 have been replaced by *Courchevel Village*, *Courchevel Moriond* and *Courchevel*, to try to break the association between altitude and level of standing (Piquerey 2016). However, at the same time, the different districts of the Chamrousse resort swapped their names of Recoin, Roche-Béranger and Bachat-Bouloud for *Chamrousse 1650*, *Chamrousse 1750* and *Chamrousse 1700* (these names do not yet appear on the official topographic maps). This double reverse movement is instructive: up to a point, the names themselves do not matter, only the decision to change them at some point to relaunch the brand, as big companies or political parties sometimes do.

In this profusion, however, care must be taken not to see neo-toponyms in all the names of tourist places: above Luchon, *Superbagneres* (Haute-Garonne) is attested from the beginning of the 17th century and so owes nothing to the much later development of the high-altitude resort (Higounet 1975). Similarly, the first writings on Mont-Blanc prove that many toponyms of the high mountains were already present before the arrival of tourists in the Chamonix valley (Gauchon 2014a).

One criterion for the success of these neo-toponyms therefore lies in their ability to endure. Not every name has the luck or the longevity of *Côte d'Azur*, either because the place itself ceases its activity, as with the *Zygofolis* leisure park which operated in Nice between 1987 and 1991, or because the name which had been promoted fades and disappears, like the *French Dolomites*, which for a time designated the eastern edge of the Vercors above the Trièves. But, on the contrary, there are rare cases where these toponyms have survived the tourist function. The snow park of *Val Pelouse* only operated for 10 years between 1976 and 1986, after which no tourist function has been maintained on site, although the French maps of the National Geographic Institute (Geoportail consultation on January 9, 2020) and road signs (autumn 2019) continue to feature it!

## 7.3. The evolution of the status and uses of toponyms: from designator to brand

The names of tourist places themselves change through the creation of neo-toponyms. But beyond toponymic production itself, the status of toponyms also changes, because tourism stakeholders see in them something other than simple designations of such and such a place. The toponym becomes a slogan or a brand and a sales pitch, and it is thus transformed according to the imagination of territorial marketing creatives and the possibilities provided by law.

What leads stakeholders in tourism to create new place names? Several reasons may be invoked, following Urry (1990), Boyer and Viallon (1994), Dann (1996) and Caiazzo et al. (2020): the new name leads to hopes for increased visibility, uniqueness or attractiveness in connection with tourist imaginations. Thus in 1966, the Chilean government renamed the two main islands of the Juan Fernandez archipelago by drawing on the repertoire of international literature: the island Mas a Tierra became *Robinson Crusoe Island*, while Mas Afuera Island was renamed *Alexander Selkirk Island*, named after the castaway who inspired Daniel Defoe (Schalansky 2010). Some names are perceived as unpleasant: when the village of La Morte (the Dead Woman), in Isère, acquired a small winter sports resort in the 1930s, commercial propriety recommended calling it *Alpe du Grand Serre*, but the village itself kept its name. Other names are not easy to pronounce, especially for foreign tourists, and their spelling is changed to remedy this, for example *La Tania*, in Tarentaise, rather than La Tagna, the name of the hamlet attested on old IGN maps.

Toponymic production also aims to designate new places such as the high-altitude resorts created in the French Alps in the 1970s. Thus, in Maurienne, after long discussions, the municipal council of Montricher-Albanne came up with the name of *Karellis* for the new winter sports resort (Lainé 2011, p. 105), according to a completely logical reasoning: new place, new name, etc. Toponymic production may also record a change of scale: for example, when in 1988 the two Pyrenean ski areas of Peyresourde (Hautes-Pyrénées) and Agudes (Haute-Garonne) were linked, the resulting new ski area was baptized with the portmanteau of *Peyragudes*. The same is true when a location changes its status: above Modane (Maurienne), the Arondaz snow stadium operated between 1969 and 1983, but when investors became interested in the site and built accommodation facilities, the new station was (re)baptized *Valfrejus*, etc. In this case, the new toponym therefore marks the transition from a place of local leisure to a place of tourism.

But beyond the status of the place, promotion of a new name aims above all to change the status of the toponym. In France, place names are in the public domain, they cannot be appropriated exclusively by particular actors, who may, however, use them for commercial activities. Furthermore, when it comes to official toponymy, as applies to local and regional authorities in particular, changes are subject to a complex procedure which involves the prefectural authorities and the Council of State, and the decisions are not always favorable to requests submitted by municipalities, departments or regions. Hence, there is temptation for economic players to attach new names to pre-existing toponyms, sometimes filed in the register of the INPI (National Institute of Industrial Property), and therefore treated similarly to brand names which are valid for a defined number of areas of activity (Gauchon 2007). The use of territorial brands therefore aims at a kind of privatization of toponymy, or more precisely, the emergence of a stratum of private toponyms which are superimposed onto public toponyms. Tourist places do not have a monopoly of this process, but use it abundantly.

In Haute-Savoie, Saint-Gervais obtained the right in 1867 to change its name to *Saint-Gervais-les-Bains* (Saint-Gervains-the Baths), which made it possible to highlight the main tourist activity of the time. But since then, the elected officials of the town have always regretted not having opted for "Saint-Gervais-Mont-Blanc" which would have ensured them greater notoriety, etc. while their neighbors in Chamonix made this winning choice in 1921. This is where a brand offers the flexibility that official toponymy does not allow: when in 2016 L'Oréal took over the management of the thermal baths and the manufacture of cosmetics, the operation was carried out under the commercial brand *Saint-Gervais-Mont-Blanc,* under which the Tourist Office has also been communicating since 2015. Without having been formalized to such an extent, the municipality of Montignac (Dordogne) communicates under the name *Montignac-Lascaux*, the suffix referring to the name of the famous cave with prehistoric paintings that is located on its territory (Giraut 2020).

Brands not only install a double toponymy but also lend themselves to a whole series of word games, in particular under the guise of "territorial brands" of which communities are increasingly fond. An anagram combined with bilingualism with the use of global English results in the formula *ONLY LYON*; the department of Isère chose in 2017 the slightly more convoluted formula *Alps is(h)ere*; and Slovenia commonly uses the spelling *sLOVEnia* to convey the subliminal message of an loveable country. The plural brand is mobilized for communications by *Perigords*, the name of the local broadcasting station for the France 3 Aquitaine network, which makes use of the well-established tourist communication around the Black, Green, White and Purple regions of Périgord, and several websites speak of the "four Périgords". Territorial actors play upon the form and the meaning of toponyms to produce new names which duplicate the official and established names. As early as 1984, the regional tourism committee of Franche-Comté foreshadowed this development by publishing a poster which extolled the advantages of *Franch'County* in a Western style.

This confusion between linguistic registers also takes advantage of the classic fluidity between toponyms and anthroponyms, as we have seen with the names of Disney parks, which attests to changes in the status and use of the names of tourist places.

## 7.4. Touristic toponymy as an element of territorial restructuring

The question of the meaning of proper names, and therefore toponyms, has already caused much ink to flow. Grevisse affirmed that "the proper name has no real meaning, no definition; it is attached to what it designates by a link that is not semantic, but by a convention that is particular to it" (1993). But not all linguists share this idea, and those involved in toponymic production are well aware that a judiciously chosen name may provide information on the amenities of the place or its territorial roots: on the Mediterranean coast, the municipalities of Beausoleil (Lovely Sun), Bormes-les-Mimosas (named Mimosas due to the proliferation of the exotic yellow blossom on the Riviera) or Palavas-les-Flots (taking its name Flots from the French word for tidal swell) offer ancient and eloquent examples.

Sometimes on the occasion of a territorial restructuring, the legislative and regulatory framework offers local and regional authorities the opportunity to change their name or create new ones to better highlight their tourist assets and produce the best combination of legitimation, promotion and development of the territories (Houssay-Holzschuch and Giraut 2008):

– Addition of an attribute relating to the tourist resource. Many municipalities have been doing this for a long time (see the earlier examples of Saint-Gervais and Chamonix), and not only in France: the two Slovenian municipalities of San Canziano and Postumia, then in Italy, were officially renamed *San Canziano della Grotta* (Royal Decree No. 800 of March 29, 1923) and *Postumia-Grotte* (Royal Decree No. 2296 of December 27, 1934), after their famous show cave system, and these names remained in force until the end of the Italian era. In another political context, Castellana (Puglia) became in 1950 *Castellana-Grotte* (Shaw 2010). In France too, from the creation of the municipality of *Le Touquet-Paris-Plage* in 1912 to the addition of Sainte-Baume to the name of the municipality of *Plan d'Aups-Sainte-Baume* in 1993, such additions were made throughout the 20th century in order to consubstantially link the tourist function to the name of the locality.

– Change of name on the occasion of the creation of a new municipality. The movement to create new municipalities, underway in France since 2015, has been accompanied by a toponymic reorganization marked by disappearances and creations of place names (Bideau and Giraut 2021; Gauchon 2021). On the whole, tourist concerns are not the reason for the majority of choices of new names, and the acceptance of the new name by the various municipalities concerned generally takes precedence. Nevertheless, certain examples are instructive: the municipality of Saint-Bon (Savoy) thus requested on several occasions to take the name of Saint-Bon-Courchevel which had been refused by the prefectural authorities, due to this designation lacking a prior history. The merger of the municipalities of Saint-Bon and La Perrière offered an opportunity to reset the toponymic clock to zero, and finally obtain the official name of *Courchevel* for the new municipality in January 2017. Several new municipalities have thus officially taken the name of the winter sports resorts which they service, such as *Les Deux-Alpes* (Isère) or *Val-Cenis* (Savoy).

The regional reform in France in 2015 did not directly lead to toponymic innovations motivated by tourism. Only the Centre region, whose boundaries did not change, took the opportunity to obtain the long-coveted name of *Centre-Val-de-Loire*, while the Hauts-de-France region, born from the merger of two regions, took the opportunity to remove the reference to the North, considered a priori unattractive from a tourist point of view. However, the avatars of some names of French politico-administrative regions illustrate in a caricatured way the entanglement of functions, interests and actors involved. The Provence-Alpes-Côte d'Azur region, whose boundaries did not change during a recent reform, officially kept its old name, which had long been criticized for its multiple spatial references, but added to it in June 2018 the brand "South Region", which now appears on the region's logo. Moreover, the former Midi-Pyrénées-Languedoc-Roussillon region chose in 2016 the official name of *Occitania*, but in the months that followed, supplemented this official name with a "signature", *Pyrenees-Mediterranean* (too bad for the Lot and Aveyron departments which are in neither of these areas), clearly identifiable in the register of tourist destinations. At the same time, the region uses the brand *South of France* to label local productions. We can thus see how tourism interferes with other concerns to lead to a production of place names which is redundant and sometimes difficult to read.

– Toponymic redundancies of communities of municipalities in France. The enforced creation of powerful entities for cooperation between municipalities since the end of the 1990s has not always been accompanied by great imagination in terms of toponymic creation. On the contrary, it has favored windfall situations where municipalities have been able to appropriate elements of tourist attractiveness through the names chosen; for example, the *Community of Municipalities of the Sources of Lake Annecy* extends to the south of the Haute-Savoie department, including towns from which you cannot see the lake. Naturalistic references, like heritage references (Bailly 2008), have often been placed at the service of this neo-toponymy, aiming both to give substance to a new web of territories and to promote its tourist assets.

In the mid-2000s, the three inter-municipal structures in the south of the Ardèche department all sought to highlight in their names the presence of the Ardèche gorges, the region's main tourist site: *Community of Municipalities from the Rhône to the Gorges de l'Ardèche*; *Community of Municipalities of the Gorges de l'Ardèche, land of men, stone and water* and *Community of Municipalities of the Great Sites of the Gorges of the Ardèche*, so that the names chosen did not make it easy to distinguish between the three new territorial entities (Duval 2007, Figure IV.25, p. 365). Since then, the map of intermunicipalities has evolved somewhat: the *Community of Municipalities from the Rhône to the Gorges de l'Ardèche* has persisted, while the other two merged into one *Community of Municipalities of the Gorges de l'Ardèche*, which hardly simplifies the toponymic landscape.

Beyond creation in the strict sense, we also observe the effects of toponymic promotion when one place name benefits from greater exposure while others are relegated to the background. These effects are particularly clear when, by metonymy, that is, the part for the whole, the space to which a place name applies is significantly extended, as in the case of several regional natural parks (PNRs). Here again for reasons of notoriety, the Vercors or Grands Causses PNRs have extended their prestigious names over less easily identified tourist areas such as, respectively, Royans or Lévezou. Since the beginning of the 2010s, the difficult gestation of a *Maritime Picardy PNR* has stumbled, among other things, on the admissibility of a flattering name to designate a space extending far inland. This toponymic dimension is probably underestimated in explaining the success of the PNR formula in France and their propensity to always extend further at a spatial level.

The same promotional effect is found for the Calanques National Park which, to the east of the municipality of Cassis, extends to the Soubeyran cliffs (municipality of La Ciotat). It is here that the idea of subversion of places by tourism joins the notion of toponymic attractor. Jean Giono, in his time, had already sensed it: "Like an oil slick, Provence overflows its historical borders" (Giono 1958, p. 21), so much so that in 2015 a *Baronnies Provençales Regional Nature Park* was created, in a space that was never part of Provence (Gauchon 2014b).

## 7.5. Conclusion

A certain form of toponymic proliferation around tourism and leisure areas leads in some cases to the coexistence of several categories of names to designate the same places: official toponymy duly consecrated by the administrative authorities; commercial toponymy established through a trademark system; informal toponymy of more or less ephemeral use. This works at different levels, and we have seen in the French case an investment in toponymy by different actors for the purpose of improving the commercial value of sites and tourist resorts, and also the way in which toponymic engineering can work at the regional level of naming.

Tourist toponymy is therefore a political affair. In general, it does not appear controversial; it is hardly the subject of heated debate, but it resonates with political intentions at different territorial levels, and political actors often appropriate the place names produced by tourism, which are enshrined and reputed to be flattering. Therefore, to return to the Côte d'Azur, Nice's airport bears the name of *Nice-Côte d'Azur*, which shows a certain logic since it is a question of serving a tourist destination. But since 2011, a metropolitan entity has existed under the same name of *Nice-Côte d'Azur*, which, on the one hand, generates possible confusion, but above all considerably extends the space of reference well beyond the coast by incorporating municipalities of the Mercantour National Park located very far in the hinterland. The race for reference to the Côte d'Azur does not stop there: the creation in 2012 of a *Regional Natural Park of the Préalpes d'Azur* managed to attach the hinterland of Grasse to the destiny of the Côte d'Azur, thus confirming the toponymic attractiveness of this appellation, of which the writer Jean Giono said that "the department manager who invented it [...] had a genius for mediocrity" (Giono 1953, p. 226).

The toponymic consequences of the establishment of tourism therefore go well beyond the places of tourism themselves and may reshape the toponymy of vast regions.

## 7.6. References

Bailly, G. (2008). Nommer les espaces de coopération intercommunale. *L'Espace politique*, 5, 79–96. DOI: https://doi.org/10.4000/espacepolitique.317.

Bideau, G. and Giraut, F. (2022). Faire territoire : logiques de la dénomination des communes nouvelles françaises. *L'Espace politique*, 43, DOI: https://doi.org/10.4000/espacepolitique.9898.

Boyer, N. and Viallon, P. (1994). *La communication touristique*. Presses universitaires de France, Paris.

Braudy, L. (2011). *The Hollywood Sign: Fantasy and Reality of an American Icon*. Yale University Press, New Haven.

Caiazzo, L., Coates, R., Azaryahu M. (eds) (2020). *Naming, Identity and Tourism*. Cambridge Scholars Publishing, Cambridge.

Dann, G. (1996). *The Language of Tourism: A Sociolinguistic Perspective*. CAB International, Wallingford.

Debarbieux, B. and Gumuchian, H. (1987a). Représentations spatiales et dénomination des territoires : l'inscription toponymique des aménagements touristiques récents dans les Alpes du Nord (Savoie). *Revue de Géographie alpine*, 2, 171–182.

Debarbieux, B. and Gumuchian, H. (1987b). Les lieux et des noms : pratiques, représentations et dénominations en haute montagne nord-alpine. In *Imaginaires de la haute montagne, doc. d'ethnologie régionale*, Joutard, P. and Majastre, J.-O. (eds). Centre alpin et rhodanien d'ethnologie, California.

Dehoorne, O. (2006). Le tourisme dans les Caraïbes, logiques régionales et enjeux environnementaux. In *La Caraïbe, données environnementales*, Burac, M. (ed.). Géode Caraïbe, Karthala.

Duval, M. (2007). Dynamiques spatiales et enjeux territoriaux des processus de patrimonialisation et de développement touristique : étude comparée des gorges de l'Ardèche et du Karst slovène. Geography Thesis, Université de Savoie, Chambéry.

Équipe MIT. (2002). *Tourismes 1, Lieux communs*. Paris, Belin.

Gauchon, C. (2007). Territoires touristiques et néo-toponymie : les noms des grands domaines skiables des Alpes françaises. *Loisir et société*, 30(1), 69–88.

Gauchon, C. (2014a). Construction toponymique ou mise en ordre du paysage touristique ? Le cas du massif du Mont Blanc. In *Paysages, patrimoine et identité*, Bouillon, D. (ed.). Éditions du CTHS, Aubervilliers.

Gauchon, C. (2014b). L'innovation toponymique comme modalité de labellisation des territoires, l'exemple du PNR des Baronnies provençales. In *Labellisation et mise en marque des territoires*, Fournier, M. (ed.). Ceramac, Clermont-Ferrand.

Gauchon. C. (2021). Des nouveaux noms pour les communes nouvelles : la production toponymique au service de la fabrique des territoires. In *Territoires fabriqués – territoires instrumentalisés : entre considération et négation de la complexité territoriale*. Bougeard Delfosse, C. et al. (eds.). Rennes, Presses universitaires de Rennes.

Giono, J. (1953). *Arcadie... Arcadie...* Gallimard, Paris.

Giono, J. (1958). *Provence*. Gallimard, Paris.

Giraut, F. (2020). Lascaux : du lieu-dit à la marque globale [Online]. Available at: https://neotopo.hypotheses.org/2385.

Grevisse, M. (1993). *Le bon usage : grammaire française*. Duculot, Paris.

Higounet, C. (ed.) (1975). Note de topographie pyrénéenne. In *Paysages et villages neufs du Moyen Âge*. Fédération historique du Sud-Ouest, Pessac.

Houssay-Holzschuch, M. and Giraut, F. (2008). Néotoponymie : formes et enjeux de la dénomination des territoires émergents. *L'Espace politique*, 5. DOI: https://journals.openedition.org/espacepolitique/161.

Kaikati, J.G. and Kaikati, A.M. (2003). A rose by any other name: Rebranding campaigns that work. *Journal of Business Strategy*, 24(6), 17–23.

Knafou, R., Bruston, M., Deprest, F., Duhamel, P., Gay J.-C., Sacareau, I. (1997). Une approche géographique du tourisme. *L'espace géographique*, 3, 193–204.

Lainé, P. (2011). *Les Karellis, de l'utopie à la réalité*. ADOPL.

Light, D. (2014). Tourism and toponymy: Commodifying and consuming place names. *Tourism Geographies*, 16(1), 141–156.

de Maupassant, G. (1886). *Mont-Oriol*. Victor Havard, Paris.

MacCannell, D. (1989). *The Tourist: A New Theory of the Leisure Class*. Schocken, New York.

Ozouf, M. (2009). *Composition française*. Gallimard, Paris.

Piquerey, L. (2016). Golden Snow. Ségrégation et entre-soi dans les stations de sports d'hiver haut de gamme en Autriche, en France et en Suisse. Geography Thesis, Université Savoie Mont Blanc, Chambéry.

Samivel (1967). *Le fou d'Édenberg*. Albin Michel, Paris.

Schalansky, J. (2010). *Atlas des îles abandonnées*. Arthaud, Paris.

Shaw, T. (2010). *Aspects of the History of the Slovene Karst 1545–2008*. ZRC Publishing, Ljubljana.

Urry J. (1990). *The Tourist Gaze: Leisure and Travel in Contemporary Societies*. Sage, London.

Urry, J. and Larsen, J. (2011). *The Tourist Gaze 3.0*. Sage, London.

# 8

# Transport Toponymy: For a Critical Study of the Toponomy of Places of Mobility

**Lucas DESTREM**
*Independent Geographer, Limoges, France*

**Figure 8.1.** *Rosa Parks station, Paris tramway. © Photograph by the author, 2015*

*The Politics of Place Naming*,
coordinated by Frédéric GIRAUT and Myriam HOUSSAY-HOLZSCHUCH. © ISTE Ltd 2022.

## 8.1. A significant but still understudied fact

The turn of the 21st century saw an intellectual impetus to renew academic thought on toponymy, and to move beyond a primarily encyclopedic treatment which had left the leadership of the discipline to archaeologists and linguists. This "critical turn" made it possible to assess a number of situations in the light of a new theoretical apparatus, and to reveal the diversity of practice and its political implications. Thus, it was possible to deal with different toponymic corpuses – odonymy (the set of names of streets and places) figuring prominently – over multiple fields, involving a number of *technologies* depending on the *actors* involved and the geopolitical *contexts* (Giraut and Houssay-Holzschuch 2016).

The toponymy that is produced for places and facilities of mobility is part of a massive, dynamic and plural spatial and political phenomenon.

This is evidenced not only by the intensification of mobilities in the context of globalization and metropolization, and their renewal at the time of energy transition, but by the increase in power (and competence) of the actors organizing them, whether local governments, private actors involved in their construction and financing, or users. The result is a significant production of territorial objects (train stations, airports, etc.), and a fortiori of toponymic landscapes linked to this context of *emergence*, which it seems interesting to question. Thus, in 2020, the world has no less than 180 metro networks, of which about a third have opened in the last decade, and to which around 30 new systems are to be added in the next two years. We might add several hundred tram networks, thousands of airports, countless private or informal public transport networks, etc. At the same time, places and objects of transport are the subject of massive symbolic and political appropriations (Destrem 2017).

The toponymy of transport thus appears to be an important field that allows spatial actors to engage in writing their landscape of signs and symbols, their *city text* (Azaryahu 1996), and to thereby perpetuate a toponymic reality in their own image, not only by creating new places but also by renaming some of them. All these places, from the simple bus stop to the international airport, each constitute hotspots at their own level from the point of view of their use or frequentation, and their role as spatial landmarks.

This toponymy remains a subject as yet little explored by the academic world. One piece of evidence of this is the absence of a term in widespread use to account for this reality. The proposal of "stathmonym" (Room 1996) has not met a response commensurate with its factual reality, even though we propose to use it here[1]. Let us also consider the boundaries of the corpus: here it includes, on an exploratory basis, the geographical places through which the use of transport networks is performed and measured, and therefore the traffic, that is, the stops and stations of rail networks, both urban and conventional, as well as airports. We will also see that this corpus might be extended, in terms of temporalities, spatial objects and geographical terrains (lines, informal networks, traffic aside from public transport, stations and stops no longer used, etc.).

To date, therefore, studies specifically centered on this "stathmonymy" and its issues remain rare, probably limited to a dozen at most, which we will discuss and list in the bibliography. It must be said that the subject has some characteristics that may have led to limited perceptions of its potential.

### 8.1.1. *A toponymy of apparent automaticity, dominated by principles of practicality*

Except in rare cases, transport networks are located in territories that are already inhabited, sometimes densely. They are expected to facilitate travel, all the more so in societies which feature an injunction to hyper-mobility, where GPS technology has permeated everything. Their toponymy therefore is largely based on an already well-named geography, resulting from the principle of "*secondary nomination*" (Bagoly-Simó and Lehmann 2017). The names of these places are in fact very largely based on pre-existing and nearby toponyms. The names of underground stations are invested with an even more imperative role of wayfinding: given the loss of awareness of the landscape and of the usual urban landmarks, names need more than anything to be practical tools linked to the surface geography. This observation also applies to main railway lines with stops that are sometimes very far apart, generating tunnel effects that may alter awareness of the territories that have been crossed.

As a result, the rhetoric of practical names dominates in official documentation, with selection criteria based on the sole objective desire to communicate the surrounding geography, as long as duplicates are avoided within transport networks[2]. Moreover, a significant number of observed name changes seem to follow the need to synchronize the names of stations to changes in the urban

1. Based on the Greek word σταθμός, stathmós (stop, station).

2. See, for example, the policy of the Geographical Names Board of New South Wales [Online]. Available at: https://www.gnb.nsw.gov.au/__data/assets/pdf_file/0017/220148/GNB_Place_Naming_Policy.pdf [Accessed 22 March 2020].

landscape, and to prevent travelers becoming lost due to ambiguous names (this is the case for almost all of the changes observed in the names of Paris metro stations since the 1950s[3]). In addition, many writings show an interest in presenting the origins of station names, and even find publishing success, but follow an encyclopedic and anecdotal, uncritical approach. Thus, like the whole of the toponymic field (Rose-Redwood 2011), the toponymy of transport seems to have been approached with a strong form of "*political innocence*". But commentary on the toponymy of transport has increased in nuance and acuity. Whether through general writings on critical toponymy (Berg and Vuolteenaho 2009; Rose-Redwood et al. 2010; Rose-Redwood 2011; Rose-Redwood and Alderman 2011), through the mainstream press[4] or even pioneering university publications, several authors have accordingly committed to going beyond preconceived ideas.

### 8.1.2. *Promising examples*

These writings have made it possible to update elements leading in the direction of a toponymy devoid of innocence and responding to the principle of the Foucauldian "spatial *dispositif*" (Giraut and Houssay-Holzschuch 2016; Rose-Redwood et al. 2018), in some places emphasizing the symbolic breadth of denominations, in others recalling the strong involvement of particular interests.

The political dimension of stathmonyms, beyond their practical function, has mainly been prominent in contexts marked by the succession of regimes with expressive and authoritarian ideologies, referring here to the contexts of *conquest* and of *revolution*. This is the case, in particular, of the cities of Berlin (Merrill 2014, 2015; Bagoly-Simó and Lehmann 2017), Moscow (Glazkov and Gnilorybov 2017; Sorokino 2020) or Kyiv (Fomenko 2018), where the significant involvement has been emphasized, first of totalitarianisms (Nazi, Soviet and aligned), then of the democratic authorities which succeeded them, in the determination of names and their subsequent changes. This results in marked *city texts*, which show the conflictuality of tormented urban histories. Thus, one assumed desire of the Nazis was to use subway names to accelerate their struggle of memory against references to the Weimar Republic (Merrill 2015). Recent observations show that these tensions also exist outside revolutionary contexts, as in France (Destrem 2015;

---

3. We may cite, for example, Rue Montmartre becoming Grands Boulevards to limit the confusion of foreign tourists thinking they were arriving at the Sacré-Cœur Basilica (1998).

4. See, for example, John Elledge on *CityMetric*, website of the British weekly *New Statesman*.

2016a, 2017), Montreal (2018), Athens[5] or in India (Kapur 2019; Adami 2020), depending on local or national ideological controversies.

The names of airports, which in some cases refer to great political figures in obvious activist tributes, also respond to these logics, and may echo important political and diplomatic issues. In their pioneering statistical study, based on a substantial corpus of 1562 sites, N. Halpern and U.K. Regmi (2011) sketch out a political reflection, then clearly distinguish a set of European infrastructures whose names are "branded", and therefore diverge from a simple desire for objective identification. Commenting on this work, some (Centre for Aviation 2015) have in turn noted the political dimension of certain denominations, underlining how issues of reputation and "pride" might guide certain choices. Besides Mother Teresa in Tiranë or John Lennon in Liverpool, we might also think of Madeira airport, which since 2017 has borne the name of the football media star Cristiano Ronaldo. In Orbán's very nationalist Hungary, the international airport has been renamed Ferenc Liszt; note the use of the Hungarian first name (Ferenc) of the composer Franz Liszt, to name this international airport after a European figure born in present-day Austria, who always used the first name Franz in his life as an artist. He is thus clearly linked to the Hungarian homeland as a great national and European figure. Mentioned several times and in another context, the case of Baghdad airport (Woznicki 2003; Rose-Redwood et al. 2010) is emblematic of the concern of certain powers – in this case the United States, through the US Air Force – to assign an ideological function to the airport, by giving it the English and global term "International", in place of the ousted leader Saddam Hussein. Finally, the case of the new Istanbul airport, opened in 2019 under this name, at the time of the creation of a major intercontinental hub, demotes in passing the Kemalist reference of the former Atatürk airport, to promote a neutral Turkish and Ottoman reference.

### 8.1.3. *The toponymy of transport: a margin of research on the political geography of mobility?*

However, studies that address the issue of station or infrastructure names show limits to their scope. For many of them, toponymy remains marginal in the study of the political meaning of transport. For example, in 2009, H. Bergmann wrote about the relationship between the railway industry and the toponyms of Central Europe, but from a primarily practical angle (inconsistencies between the terrain and the administrative names, duplicates, identification of landscape referents involved in

---

5. Keep Taking Greece (2019). Greeks disapprove of the Athens Metro decision to rename two stations [Online]. Available at: https://www.keeptalkinggreece.com/2019/09/21/athens-metro-rename-stations/ [Accessed 29 March 2020].

the choice). He scarcely recognized the use of certain names to enhance the tourist visibility of a site or locality of less importance than where the station is located. Halpern and Regmi (2011) also make only limited use of the political question, sticking to an inventory of homages and an allusion to the risks of divisions that these choices may generate. The authors also recognize that pursuing a more qualitative investigation would be appropriate.

Other studies present cases where the political function is established, but remains correlated to a previous production of names, largely odonymic. This is the case of the Mexican stations studied by A. Musset (2010), whose names are more a "reflection" than a real "engine" of the toponymic political project. We also encounter, in part, this limit in several French cases (Destrem 2016a, 2017), and in that of Brussels studied by Frédéric Dobruszkes (2010), which must concede a rather weak political motivation in the choices. The great majority of the names are a tracing of the landscape – names of streets or infrastructures – enacted without clashes or media debates. Thus, the political dimension of station names seems at several points underestimated, not very visible or quite simply difficult to grasp because of their association with pre-existing geographical objects and names, or description of landscape realities. This is shown by Gin and Perono Cacciafoco (2020) for the Singapore metro, where the political dimension is more apparent in the languages that are chosen.

Another limitation of the existing corpus is that the main explicitly political case studies focus on emblematic, "impressive" cases, those of *conquest* and *revolution*, such as Berlin, Baghdad, Moscow or post-apartheid South Africa, where several airports, for example, have also lost their homage to the tutelary figures of Afrikaner nationalism (Giraut et al. 2008). These telling cases open up fundamental and stimulating debates, but they nevertheless also require dealing with many fields which are perhaps less spectacular, and also more recent.

A first group of work engages with this diversification, without losing sight of the political purpose. This is the case of Olena Fomenko (2018), who is interested in Kyiv metro stations. She studies 21st century modifications, which far from becoming free from the weight of Soviet memory, are integrated into the thwarted Europhile discourse of contemporary Ukraine. For example, "Olimpiiska" takes the name of the stadium, and since 2011 has replaced a tribute to the Republic with a not-so-paradoxical double allusion to the 1980 Moscow Olympics and the holding of the 2012 European Football Championship, which echoes the global influence of the Games. In France, in a recent comparative study, tram lines appear to be a major tool for political communication and justification. Far from bloody revolutions, a very large majority of stations (98%) clearly relate to the wise principle of "*secondary nomination*", which does not prevent the assumed communication of strong political messages such as the quest for metropolitan influence or the affirmation of intermunicipalities (Destrem 2017).

These studies contribute to updating the theoretical framework and lines of research around current issues, as perceived elsewhere in critical geography, and provide an overview of trends to explore, or hypotheses to confirm or to nuance. We will come back to it.

### 8.1.4. *The commodification of names: various practices that shed light on the motivation of politics*

Alongside revolutionary changes, interest in transport toponymy has also focused on the commodification of names and the phenomenon of the sale of naming rights, which also relate to the spectacular. This practice should be understood here as relating to the signing of a contract giving to a place the name of a company for a given period, in exchange for a financial payment to the public structure that owns it (see Chapter 6 in this book).

Studies have focused on this attractive "win-win" model (new customers for firms, new resources for infrastructures), this tension between "use value" and "exchange value", in particular based on the iconic case of sports arenas. Promoted by several geographers (Rose-Redwood et al. 2010; Light and Young 2015), understanding the motivations and criteria of choice for the localization of *naming* and the study of new fields have met with a favorable response. It is through this problematic, applied to transport infrastructures, that the general public and the general press have come to take an interest in critical stathmonymy.

This phenomenon of commodification, in fact, increasingly concerns the toponymy of transport: in the Middle East (Dubai, a pioneer in 2008 with 23 of its 47 metro stations using names offered for 40 million Emirati dirhams each for 10 years); in India (Mumbai and Delhi metros in particular), in the United States (11 stations offered in Chicago, some media examples in New York with the Barclays station, etc.); in Brazil (where in January 2021, the famous station of Botafogo in Rio acquired the added name of Coca-Cola, a firm that has its Brazilian headquarters not far away); in Eastern Asia (Japan, China, Indonesia); and to a lesser extent in Europe, which constitute the main areas discussed in press articles (including those of *CityLab*, a former offshoot of the American monthly *The Atlantic*) and some scientific studies, which are only partially devoted to it (Kapur 2019; Rose-Redwood et al. 2019). Transport operators also produce enlightening documentation on their intentions in this area, in which, for example, they detail and compare contract clauses[6]. These writings now make it possible to report on the modalities and singularities of this phenomenon in the field of transport, producing inventories of practice on a global scale, informing the development of strategies of *semi-naming* (where the addition of suffixes and the *branding* of spaces are preferred to a complete name change, as in the case of the Botafogo-Coca-Cola station), as well as some legal details. Interesting cases arise, such as in Madrid with the Vodafone line, in Japan (see Figure 8.2) or in the Indian metropolises, also making it possible to place the practice in historical perspective and to report contestations and refusals (as in Boston).

**Figure 8.2.** *Sign at the Japanese station Kasagami-Kurohae, modified as part of a promotional contract with a hair products company. "Kaminoke" means "hair" in Japanese. © TRJN, 2016 (license CC BY-SA 4.0)*

6. See, for example, Hoyez, M., Ferro, P. S. (2016). Selling metro station naming rights: examples and debates. CODATU [Online]. Available at: http://www.codatu.org/actualites/selling-metro-station-naming-rights-examples-and-debates/ [Accessed 9 March 2020]; Mumbai Metro Rail Corporation Limited (2019). Expression of Interest for Semi-Naming and Branding Rights for Mumbai Metro Line 3 stations [Online]. Available at: https://www.mmrcl.com/sites/default/files/2019.12.26_Semi-Naming%20Rights%20EOI.pdf [Accessed 21 March 2020].

Using analytical tools developed in the field of sports equipment and based on the geography of neo-liberalism, the subject would benefit from studies entirely devoted to it, using a toponymic and territorial approach. The penetration of the offensive marketing practices of capitalism into the functioning of public organizations is a promising angle. Some have documented cases of "public-public" *naming*, as in Denver, where the university has sponsored and "named" a train line that does not serve it (see Figure 8.3)[7].

Commodification practices not only cover cases of surrendering toponymic competence, but also the promotion of names as advertising tools. Without going as far as the transfer of ownership, but rather by exerting pressure of the *soft power* variety at the local level (strong media presence, for example), other public structures, such as the universities of Helsinki (Lehtonen et al. 2014) or the French intermunicipalities for their urban projects (Destrem 2017), manage to have "their" names erected into stathmonyms. From then on, these function as calling cards. In this way, these structures don the habits of competitiveness and competition, and turn their territorial influence or their democratic legitimacy into arguments of authority asserted by their executive (university presidency, presidency of a public establishment for inter-municipal cooperation).

**Figure 8.3.** *Inauguration of the "University of Colorado A Line" at Denver (United States) in 2016. © Xnatedawgx (license CC BY-SA 4.0)*

7. Hoyez, M., Ferro, P. S. (2016). Selling metro station naming rights: examples and debates. CODATU [Online]. Available at: http://www.codatu.org/actualites/selling-metro-station-naming-rights-examples-and-debates/ [Accessed 9 March 2020].

This type of subtlety, vital to detect and reveal, allows us by means of an economic and financial approach to glimpse the extraordinary political significance of stathmonymy. There are many of these, including in studies already produced. These signals invite us to abandon "political innocence".

### 8.1.5. *Dealing with geography: a toponymy meaningful through its stratagems*

Very often squeezed into already strongly named urban spaces (except in the case of peri-urban settler frontiers), transport networks appear to offer toponymic opportunities, and their public decision-makers compete with stratagems to satisfy their desire for new commemorations without resorting to a cleansing of odonymy. They achieve this, through various stratagems or arrangements vis-à-vis the sacrosanct idea of stathmonyms naively inherited from the toponymy already in place.

These actors may play with the notion of distance, by giving the station the name of an urban landmark that is actually nearby, but perhaps further away than other references that will be neglected, for lack of the required ideological dimension. In Paris, to respect the political decree of the names of tram stations promoting famous women of the world – in the name of the equalitarian and internationalist policy of the city (see Figure 8.1) – the municipality, in conjunction with the public company that operates the network (RATP), has selected female odonyms that are sometimes several hundred meters away (Destrem 2016a). Giving the name of a locality to a station that is not actually there is a recurrent practice: the Crainhem station, in the Brussels metro, promotes the nearby Flemish municipality, while remaining on the territory of the bilingual region of Brussels-Capital, a reward for the lobbying of a Dutch-speaking minister (Dobruszkes 2010). This phenomenon is also observed at certain tram terminals such as Brest, to confirm the metropolitan financing of a project which is nevertheless limited to three municipalities. To ensure their attractiveness, many secondary airports are also keen to attach the name of an important but distant city to their main point of reference (the world record goes to Paris-Vatry, 160 km from the French capital). The same goes for TGV stations referring to an encompassing territory[8]. Some may compromise with the notion of "urban landmark" to legitimize the station's name: in Athens, the location of the assassinations of honored personalities; in Paris, their place of life (Alexandra David-Néel) or of burial (Barbara). Decision-makers may also play with the notion of time: for the name to refer to a pre-existing geographical landmark, we can

8. The TGV stations have also introduced novelty and debate through their terminology that smacks of place branding (Valence-TGV Sud Rhône-Alpes, Nîmes-Pont du Gard).

"force" the coincidence, by hastening to create the said landmark before the station rises from the ground. This is what was observed in Paris, between 2012 and 2013, where the streets Delphine Seyrig and Ella Fitzgerald were named during the construction works of the tram stations which were supposed to then take those names.

It is also possible to play with the names themselves, without departing from the rules: for example, by adopting subtitles, which instill homage without disrupting uses and requiring too costly investments: Europe – Simone Veil in Paris (Charbonneaux 2020); the stations of the Corsican railway network accompanied by their translation into the island language, the signature of Dominique Baudis on the pediment of Toulouse airport… rather than rebaptism.

Finally, let us recall that the simple reiteration of nearby names, largely the majority of cases, may not exclude political intentions! These borrowings are very often magnifying glasses for the odonymic *city text*, which thus reinforces the visibility of figures, ideas and events in which regimes are the actors. The major French cities are therefore keen to place the infrastructures of metropolization (theatres, hospitals, stadiums), of which they were the originators, into station names, under cover of practical intentions. Born from overflowing political ambition, some cases, the extent of which remains to be measured and of which I have given examples for French tramways (Destrem 2017), even seem to free themselves from any pre-existing landmark, such as the Martin Luther King station of Orleans. In a rather original reversal, some examples also show a toponymy of transport that gives rise to odonymy. In Berlin, S. Merril (2015) recalls that changing the names of certain stations preceded and facilitated the changing of street names (Horst-Wessel Platz under Hitler; Thälmannplatz at the beginning of the GDR[9]).

The practices, objectives and strategies are therefore numerous, and the naming processes provide at least as much information as the names themselves.

## 8.2. Research perspectives

Several projects have been assigned to emerging critical toponymy studies (Rose-Redwood et al. 2010; Rose-Redwood 2011; Rose-Redwood and Alderman 2011; Light and Young 2015). The hindsight we now have on recent works allows us to judge their application to the subfield of stathmonyms.

9. These references (street as station) then replaced respectively "Schönhauser Platz" (former historic enclosure) in 1934 and "Kaiserhof" (former hotel) in 1950.

### 8.2.1. *Documenting to update*

The recent nature of the geographical interest in stathmonyms and the concealment of their political dimension have not facilitated knowledge of the corpus, notably the contemporary corpus stemming from recent policies for the development of new lines, particularly those under strictly private sector initiative in the Global South and its informal settlements. The identification of the traces is, however, necessary. By *traces*, we mean all the documentary elements which, in advance, clarify the debates – rarely public, for the most part – the motivations, the possible conflicts which result in the denominations, the names which are envisaged and recalibrated – no less significant!– and the identity of the actors involved. The aim here is to question the competent services and qualified witnesses, to access any archives… and to convince the actors in charge of toponymy to leave archives behind (unless their rarity is deliberate?).

The creation of an evolving observatory of the territoriality of mobility, and therefore of the production of toponyms attached to it, on a global scale, would thus be welcome. Initially, this panorama would already be useful for updating definitions and typologies, measuring the phenomenon and identifying some trends to be contextualized (like those involving airports). Consequently, spatially differentiated studies would be undertaken, applied to the theoretical framework set out recently, on the analysis of changes, renunciations, the formality of processes.

### 8.2.2. *The question of scale in the age of "glocal" mobility*

The question of scale (Hagen 2011) deserves to be addressed here in several ways. The first angle is that of opening up the reflective framework to new geographical and sub-thematic terrains (we will conclude with this point). The second angle is that of the scale of references of toponymic signifiers. As geographical places of intense mobility, airports and stations refer as much to local realities (the region, the street they serve) as to global ones (a world-city, a leader, a concept). Airports are a very stimulating field. Their toponymy inscribes them in multiple territorialities: global (through the English word "International", the fact that duplicate names on a planetary scale are scarcely thinkable), national (the "capital" airport honors a national hero, sometimes little known outside the borders as is the case with Adolfo Suárez in Madrid or Jomo Kenyatta in Nairobi[10]) and local (see the case of Glasgow Airport's slogan, "*pure dead brilliant*", evoking "the best" in Scottish slang, but whose literal translation, taken out of context, has caused controversy). In general, transportation is liberated from administrative areas and

10. Respectively first democratic head of government and first president of the country.

schemes, and therefore transgresses the limits of legal and ideological corpuses. Hyper-places of globalization and symbol-places of the State, airports adjoin railway networks which may also cross borders and question the responsibility for their toponymy. Hence, a third angle: who decides, at which levels?

### 8.2.3. *Interrogating skills to read intentions*

The effort of documentation must therefore also concern the engineering, the models of "toponymic governance", whether they are thought out or not. It is an indispensable condition for understanding intentions, and thus, perhaps, deciphering denominations which are a priori perceived as obscure or poor, which they are not.

Early works, already cited, have made it possible to underline the importance of interplay between actors in the extent of the ideological meaning of denominations. Some indications emerge, on the primacy of the responsibility of institutional actors, in dictatorships (leaders) as in democracies (mayors), or on the secrecy of procedures, generally little promoted by decision-makers who consider them secondary, even though participatory initiatives are to be watched carefully. These initial reflections raise clues (debates, internal regulations, personal balance of power, interests, etc.), which shed light on political projections in the toponym. It is therefore crucial to generalize the effort to seek in-depth knowledge of the personnel involved and of the creative processes, in particular the most contemporary ones, in order to learn all the ins and outs. A tightening focus at the individual level, as advocated by Light and Young (2018), seems essential. In particular, this involves questioning the distribution of roles, both regulatory and practical, between the three types of actors that have been theorized (governments, private sphere, civil society). If cases of conflict are numerous, more nuanced and homogeneous situations are to be targeted no less. The practices of the private sector, which as in the case of Algiers may exceed their role without a contractual framework (Atoui 2014) or which may offer assistance to politics by engaging with the field of ideas[11], shake up the institutional sphere, in a more subtle way but not necessarily less effectively than in the case of direct naming. In the latter case, the examples of the networks concerned are, of course, interesting: the exploration of commodification practices should be extended (studying unsuccessful cases, emerging trends, undertaking comparative analyses, etc.).

11. While waiting for the opening of new metro stations in Paris, the women's ready-to-wear brand Un Jour Ailleurs has invited Internet users to suggest names of women via the site desfemmesasuivre.com.

Equally important is the consideration to be given to conciliation within groups of actors. Within the governmental sphere, the degree of technocracy must be questioned, starting from the first teachings that are read here and there. How is the balance established between the culture of services and the political culture of elected officials? What is the impact of these relationships on toponymic choices (in terms of responsibility and final choice)? Are commemorative names always the prerogative of politics? Are descriptive names that of technical services and administrations? We have been able to underline for stathmonymy the extent to which technical culture might, in certain cases, become very widely infused into regulations and public action, unlike other fields where ideological or even activist discourse remains very strong (Destrem 2016b). We are also beginning to see the extent to which technical constraints and toponymic strategies may serve each other, and generate not-so-trivial toponymic results (requiring a short name, for example, generates slogan-names). All this without neglecting the idea that apology for technology remains a very ideological discourse! Relations between municipal and nationally elected officials (as approached by S. Merrill in the case of post-1989 Berlin) are another line of research.

Particular interest might be given to other areas, little covered or not at all:

– the effective role of commissions of toponymic experts;

– the processing of these names in databases or in national and international standardization procedures;

– the evocation of stathmonymic competence (creation and management) in legal texts, and in particular in certain contexts generating complexity (cross-border networks and transnational transport companies, decentralized or federal systems, open borders, etc.).

In the same way that we question what is behind the absence of changes or of spectacular names, it is also essential to dissect the contexts where no established strategy seems to structure the production of stathmonyms, and where official discourses are constrained by the informality of the networks and the dilution of transport options. The urban areas studied by C. Cumbe in Maputo (2016) and H. Yaye Saidou in Niamey (2018) are marked by the great spontaneity of unofficial, oral stathmonyms linked to private networks of public transport. These toponyms of use, mixtures of different languages and references to public and private, formal and informal, official and vernacular places, are nonetheless subject to issues of power. We can question the dynamics that pass through them, their relationship to institutionalization, not to mention their practical function and identity.

Finally, questioning the dynamics of actors requires questioning the place of inhabitants and users in the toponymic practice of places of transport. Rose-Redwood et al. (2010) have already called for interrogation of the power struggles around toponymy. This questions, in particular, the effective participation, or otherwise, of the population in the determination and use of stathmonyms. The press and certain academic works echo popular demands or institutional initiatives inviting citizens to give their opinion, more or less autonomously, on the choice of station names, as in Singapore, Melbourne or Brest where various consultation processes through online voting have been put in place. But in general, the discipline would benefit from better questioning the skills and expectations of "ordinary citizens" in the case studies, whether they are absent, passive players or really involved. Including on the question of the impact of names on day-to-day social relationships with space, or on the proven use of official toponymies by users, a subject for which the study of names of informal mobility, for example, is to be pursued (Dorier and Van Den Avenne 2002).

### 8.2.4. *Social and spatial impacts of stathmonyms*

Glazkov and Gnilorybov (2017) suppose that the Moscow metro has an impact on the city, through its primary function (real estate prices, frequentation of neighborhoods) and also through its toponymy, since the inhabitants use the names of stations and lines to move around, as if to represent themselves and define their social level.

The spatial impact of stathmonyms can be measured well in several ways. First, it would be a question of questioning the effectiveness of the toponym in its role of aid for wayfinding and movement. This question, seemingly naive but difficult to deal with, conceals many others, with multiple implications that we outline below. Much remains to be done, here again, in analyzing the material and symbolic ripple effects of stathmonyms on populations and representations. Thus, as with the Proustian evocative power of toponyms in general, underground names in particular shape the imagination, the representations we have of the surface (Dukes 2018). What is the impact of the naming of lines and stations on collective and individual practices, in travel and also in all daily actions[12]? What connections can be established with civic mobilizations that engage with the urban space, related to gender, minorities, accessibility? The prospects are vast and numerous. The

12. These impacts become… toponymic: it was recalled that the names of stations might themselves generate names for the surrounding district (as in London where certain names are no longer based on referents but have become such), or for a street created simultaneously. Geo-historical analyses might support this supposition.

expected impact of names modified by commodification relates to other types of social impacts, those of marketing and consumerism. Analysis of individual responsibilities and behaviors, which we have already suggested for decision-makers, might materialize this ambition, if it is multidisciplinary, including the neurosciences, and involves extensive surveys of recipients and users of transport toponymy.

Interest might be given to material objects that communicate toponymy (signs, official and private maps, street furniture) which already interest researchers (Piveteau 2003; Musset 2010), and have stimulated creation by certain artists, outside of official *dispositifs* as well as within them (see, in particular, the Paris metro map in which Silvia Radelli replaced all the names with women, or the Oulipiens action on the Strasbourg tram, or even the Paris metro map that I adapted for the RATP (Figure 8.4)). Thus, it might be possible to correctly assess all the consequences of the contracts of *naming* because these consequences cannot be financial alone, even though their study remains to be carried out seriously (Capps 2019). At the same time, in fact, the expected impact of these contracts on the public finances of network operators is the subject of interpretations which do not necessarily indicate unequivocal success, and whose full accounting success is debatable[13]. This in no way dampens the ardor of decision-makers to continue its use, convinced that the impact of a "branded" station will be more significant than that of a street, in that it is more frequented, visible and verbalized by everyone. The question also remains unanswered of the potential effect of the names of airports on their commercial activity. The names of stations or stops, perhaps more visible (in particular, the terminals which give their names to the lines), more "obligatory" (it is less easy to bypass a station than a street), having a greater impact on lifestyles and imaginaries[14] than odonyms, are also a privileged medium for activist demands and media demonstrations. This would involve studying the methods of appropriation, in that they may differ from other types of toponyms. For example, what is the place of the relationship to mobility, in terms of socio-spatial justice, in the positions taken on stathmonyms?

13. See Jaffe, E. (2012). The problem with naming transit stations [Online]. Available at: https://www.bloomberg.com/news/articles/2012-01-04/the-problem-with-naming-transit-stations [Accessed 23 July 2020].

14. The case of Penny Lane, the terminus of a tram line in Liverpool, which inspired the famous Beatles song in the 1960s, illustrates the evocative power of such names. In this case, totally detached from its original signifier relating to the figure of John Penny, a slave trader of the city (Giraut 2014).

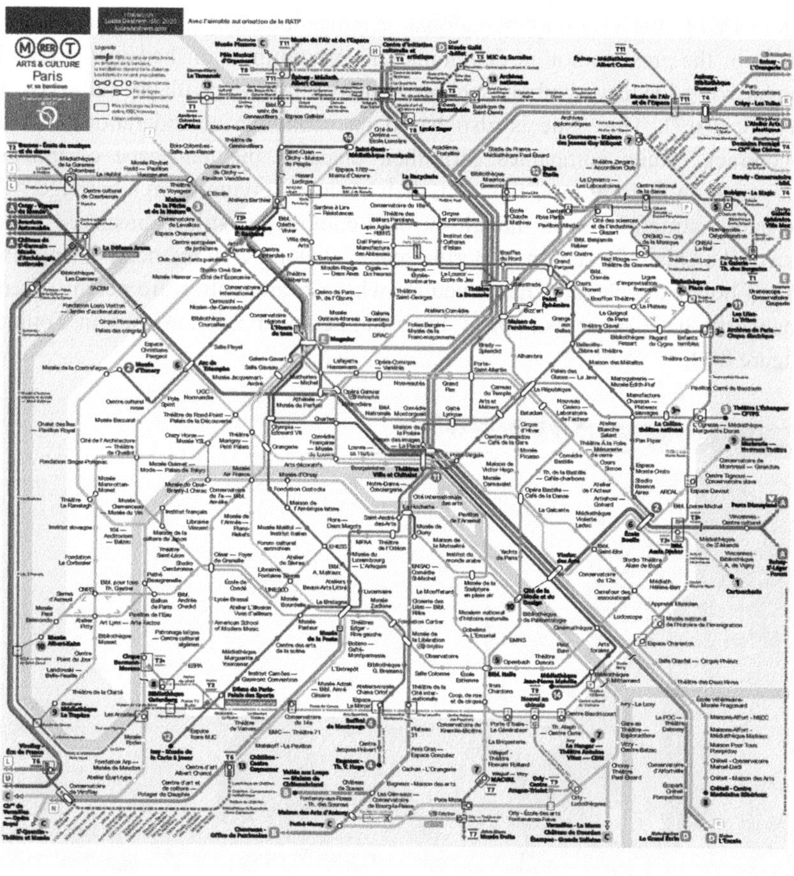


**Figure 8.4.** *"Arts and Culture" plan of the Metro, RER and Tramway network in the Paris region adapted from the official plan by Lucas Destrem for the RATP with renaming of stations based on notable cultural facilities (© 2021). For a color version of the figure, see www.iste.co.uk/giraut/naming.zip*

The heritage status of stathmonyms seems to us to constitute another relevant area of investigation for studying and understanding the social effects of this toponymy. It cannot be reduced to the heritage enhancement of toponymy and ancient networks: it also includes the appropriation of heritage objects as political foundations, tools and pretexts for action. This is evidenced by the crystallization of debates related to the visibility of regional or indigenous languages in names (as in Hawaii[15]), or the preservation of denominations promoting vanished spatial realities but emphasizing identity symbols such as the old Muscovite villages (Glazkov and Gnilorybov 2017). The issue is also based on economic considerations: without juicy contracts, some companies earn their place in the landscape of transport stations based on their cultural function, their seniority giving them a heritage value dear to public structures even though the said company is still in business! This is the case of the Otistraße station in Berlin, renamed along with the neighboring street in 2004 to honor the elevator manufacturer whose headquarters are nearby. French tram networks have other examples: Trois Suisses near Lille, Airbus in Toulouse (see Figure 8.5).

**Figure 8.5.** *Servanty-Airbus tram station at Toulouse.*
*© Photograph by the author, 2015*

15. See Honolulu Authority for Rapid Transportation (2019). Hawaiian Station Naming Program [Online]. Available at: http://hartdocs.honolulu.gov/docushare/dsweb/Get/Document-22717/HSN-Program-Overview-2019.pdf [Accessed 9 March 2020].

### 8.2.5. *Other fields to question*

In addition, reflection may be enriched by the study of peripheral land that is still poorly documented. This is the case with Asian, African, South American and generally non-English-speaking areas. Various relationships to space, institutions, or linguistic heritage must be studied to contribute to the general reflection by highlighting more informal and vernacular practices and actors. Positive perspectives are outlined by recent communications around fields of informality, in particular in Africa[16]. These early works are valuable and decisive milestones. Throughout the planet, including in Western states, a priori better known, entire regions also await the interest of researchers. Rural and peri-urban networks, if they are studied, may constitute fields capable of documenting contexts of *emergence* and technologies of *promotion*, for the moment less visible, or perhaps to diversify the profile of *revolutions* or *eradication*. Exploration might also be geo-historical, for example, analysis of bygone but probably enlightening strategies (toponyms of places of archaic mobility and communication, such as post houses or the first trams).

Let us also look at how the toponym is dressed up, how it is presented to users: its spelling, its translations revealing linguistic choices sometimes criticized as belonging to a hegemonic linguistic regime (De Los Reyes 2015); its temporality, which in the light of history, can remind us that an accepted terminology, in fact, refers to a bygone, loaded use[17]. Various visual presentations may be the subject of our attention: signage, pictograms, street furniture, which schematize or even stereotype and reinterpret certain names; event-driven renominations, which under cover of animation and connivance, may support advertising practices. The aestheticization of names (through art, for example), the marketing of derived objects, induce more global strategies of animation and promotion of urban space, of questioning and democratic inclusion, which echo other practices: the gamification of urban practice, gentrification, place branding as means of mobilization (see Figure 8.6).

16. Long limited to South Africa and Algeria, toponymic reflection extends at least from the reflection of E. Dorier and C. Van Den Avenne in Mali (2002) to the whole continent, as evidenced by the work edited by Liora Bigon (2016) or the international symposium organized in Niamey in 2018.

17. Merrill recalls that "Belle Alliance Straße", in Berlin, referred to the way the Nazis referred to the battle of Waterloo, which may be unknown, out of context. "Rosa Luxemburg" was derided there as a Marxist, but later saved as a female figure.

**Figure 8.6.** *Call for stathmonymic puns on the Lyon network, with free subscriptions to be won. Décines and Meyzieu are Lyon suburb station names; they are homophonous in French with the phrase "Draw my eyes, with just one stroke". © Photograph by the author, 2014*

Our remarks and observations are based on an ultimately incomplete corpus. What about extra-metropolitan stathmonymies? Are they subject in the same way to the relationship of tension between technicians and politicians, to pressure from the private sector? Do they pursue the same ideal of competitiveness? What about the national railway networks, whose old European specimens present a toponymic corpus that is a priori stable and prior to the emergence of local governments, but with a real heritage status?[18] What about carpooling, car-sharing and self-service bicycle areas, whose companies are evolving in the context of public–private partnerships? What can we say, finally, of the private car, and of the relationship that it, speed and signs, may undoubtedly suggest to us, drivers or planners, with regard to the toponymy encountered along the way? By devoting a documentary

18. TGV stations have also introduced novelty and debate, through their terminology that smacks of place branding (Valence-TGV Sud Rhône-Alpes). In the Swiss context, the new Greater Geneva train network (the CEVA or Cornavin-Eaux-Vives-Annemasse) has also seen municipalities fight to obtain mention of their name, rather than that of the agglomeration, in the composition of the station name.

film to the production, reception and ultimately the meaning of heritage signs on French highways, Seb Coupy (2019) opens this path as a semiologist and film-maker.

Beyond the toponymy of "places of mobility" such as stops, stations or airports, perhaps we might thus open our curiosity to the toponymy of travel and distances. How do we name lines and roads, and what political intentions do we place on them?[19] This is a stimulating reflection that some, like Piotr Tomasik (2018) (see Figure 8.7), have opened up. How are point-based toponymies and linear toponymies articulated? And what might our travels, in short our first experience of transport, induce in toponymic terms? Is walking an original driver of stathmonymies? We might think so, insofar as many bus, tram or metro stops do not infuse the same area, influencing thereby the toponymic referents that we might assign to them[20].

**Figure 8.7.** *Budapest trolleybus no. 70 owes its number to its inauguration on Stalin's 70th birthday. © Bakó Ádám, 2008 (license CC BY-SA 3.0)*

19. Let us therefore recognize that the term "stathmonymy", which we have decided to use, probably fails to characterize these non-point spatial realities (lines, networks, etc.), so that it may be discussed in the future reflections that we call for.

20. Thus, the RATP identifies around stations, according to their accessibility, these "circles of influence" that direct the toponymic choices.

We must definitely put an end to the political innocence of transport toponymy. Numerous signs and beginnings of interpretations allow us to see the extent to which stathmonymy, while being very much part of the frameworks posed by critical toponymy, is rich in specificity. It would benefit from new studies promoting its typicality, and the autonomy of some of its sources, over the actions and representations of human beings.

## 8.3. References

Adami, E. (2020). Constructing lines, constructing place-names: The case of the Delhi Metro. In *Shifting Toponymies: (Re)naming Places, (Re)shaping Identities*, Caiazzo, L. and Nick, I.M. (eds). Cambridge Scholars Publishing, Newcastle upon Tyne.

Atoui, B. (2014). La société de gestion du tramway décide des appellations : une station, cinq noms ! [Online]. Available at: https://www.elwatan.com/pages-hebdo/magazine/la-societe-de-gestion-du-tramway-decide-des-appellations-une-station-cinq-noms-27-11-2014 [Accessed 29 March 2020].

Azaryahu, M. (1996). The power of commemorative street names. *Environment and Planning D: Society and Space*, 14(3), 311–330.

Bagoly-Simó, P. and Lehmann, M. (2017). Berlin underground toponymy. *23rd International Seminar on Sea Names*, Berlin.

Berg, L.D. and Vuolteenaho, J. (2009). *Critical Toponymies: The Contested Politics of Place Naming*. Ashgate, Burlington.

Bergmann, H. (2009). Geographical names and the railway industry. *UNGEGN, 25th Session*, Nairobi, 1–10.

Capps, K. (2019). The "namewashing" of public transit [Online]. Available at: https://www.citylab.com/transportation/2019/11/washington-dc-metro-station-names-public-transit-fares-map/602290/ [Accessed 9 March 2020].

Centre for Aviation (2015). What's in an eponym? Celebrity airports – Could there be a commercial benefit in naming? [Online]. Available at: https://centreforaviation.com/analysis/reports/whats-in-an-eponym-could-there-be-a-financial-or-other-benefit-in-celebrity-airport-naming-215965 [Accessed 16 March 2020].

Charbonneaux, J. (2020). Du toponyme à l'incarnation métonymique : la renomination parisienne de la place et de la station de métro "Europe" en "Europe-Simone Veil". *De Europa*, 3(2), 153–163.

Coupy, S. (2019). L'image qu'on s'en fait. Documentary, Tenk.

Cumbe, C. (2016). Formal and informal toponymic inscriptions in maputo: Towards socio-linguistics and anthropology of street naming. In *Place Names in Africa*, Bigon, L. (ed.), Springer, Zurich.

De Los Reyes, R.A. (2015). Language of "order": English in the linguistic landscape of two major train stations in the Philippines. *Asian Journal of English Language Studies*, 2, 24–51.

Destrem, L. (2015). Les noms du tram : la toponymie des réseaux de tramway dans les aires métropolitaines françaises. Research paper, Master's Degree, Université de Genève, Geneva.

Destrem, L. (2016a). De la ville-musée à la ville-monde, les enjeux de la toponymie du tramway parisien [Online]. Available at: https://neotopo.hypotheses.org/349 [Accessed 1 March 2020].

Destrem, L. (2016b). La toponymie comme enjeu de la gouvernance transfrontalière : le cas des noms de stations du tramway du Grand Genève [Online]. Available at: https://neotopo.hypotheses.org/607 [Accessed 1 March 2020].

Destrem, L. (2017). Les noms du tram. Quand les dénominations des stations disent le projet de ville. *Annales de géographie*, 713(1), 5–30.

Dobruszkes, F. (2010). Baptiser un grand équipement urbain : pratiques et enjeux autour du nom des stations de métro à Bruxelles. *Belgeo*, 1(2), 229–240.

Dorier, E. and Van den Avenne, C. (2002). Usages toponymiques et pratiques de l'espace urbain à Mopti. *Marges Linguistiques*, 152 158.

Dudemaine, C. (2018). Toponymies politiques : les luttes pour nommer et renommer. *Bulletin d'histoire politique*, 26(2), 258–274.

Dukes, H. (2018). From Balbec to Ballyba : toponyms, transportation, and the etymological imagination [Online]. Available at: https://modernismmodernity.org/forums/posts/balbec-ballyba [Accessed 15 August 2022].

Fomenko, O. (2018). The metro and the city: The linguistic landscape of Kyiv underground and remaking national identity. *International Scientific Conference Onomastic Investigations*, Riga, 40–41.

Gin, S.L.T. and Perono Cacciafoco, F. (2020). Then and now: A comparative historical toponomastics analysis of station names in 2 of Singapore's mass rapid transit (MRT) lines. *Urban Science*, 4(3), 37.

Giraut, F. (2014). Penny Lane comme méta toponyme [Online]. Available at: https://neotopo.hypotheses.org/30.

Giraut, F. and Houssay-Holzschuch, M. (2016). Place naming as dispositif: Toward a theoretical framework. *Geopolitics*, 21, 1–21.

Giraut, F., Guyot, S., Houssay-Holzschuch, M. (2008). Enjeux de mots : les changements toponymiques sud-africains. *L'Espace géographique*, 37(2), 131–150.

Glazkov, K. and Gnilorybov, P. (2017). "Nomadic" toponymy: Urban transport system as driving force of moscow toponyms. *HSE*, 2(112), 232–252.

Hagen, J. (2011). Theorizing scale in critical place-name studies. *ACME: An International Journal for Critical Geographies*, 10(1), 23–27.

Halpern, N. and Regmi, U.K. (2011). What's in a name? Analysis of airport brand names and slogans. *Journal of Airport Management*, 6(1).

Kapur, A. (2019). *Mapping Place Names of India*. Routledge, New York.

Lehtonen, J., Mallat, K., Suviranta, S. (2014). Naming metro stations in Helsinki and Espoo. *Names and their Environment. Proceedings of the 25th International Congress of Onomastic Sciences*, Glasgow.

Light, D. and Young, C. (2015). Toponymy as commodity: Exploring the economic dimensions of urban place names. *International Journal of Urban and Regional Research*, 39(3), 435–450.

Light, D. and Young, C. (2018). The politics of toponymic continuity: The limits of change and the ongoing lives of street names. In *The Political Life of Urban Streetscapes: Naming, Politics and Place*, Rose-Redwood, R., Alderman, D., Azaryahu, M. (eds). Routledge, New York.

Merrill, S. (2014). Excavating buried memories: Mnemonic production in the railways under London and Berlin, 1933–2013. PhD Thesis, University College London.

Merrill, S. (2015). Identities in transit: The (re)connections and (re)brandings of Berlin's municipal railway infrastructure after 1989. *Journal of Historical Geography*, 50, 76–91.

Musset, A. (2010). Le plan du métro de Mexico : points, lignes et territoires. *Le monde des cartes*, 204, 77–84.

Piveteau, J. (2003). La signalisation routière de direction : une nouvelle donne dans notre relation au territoire. *L'Espace géographique*, 32(3), 193–204.

Room, A. (1996). *An Alphabetical Guide to the Language of Name Studies*. Scarecrow Press, Lanham.

Rose-Redwood, R. (2011). Rethinking the agenda of political toponymy. *ACME: An International E-Journal for Critical Geographies*, 10(1), 34–41.

Rose-Redwood, R. and Alderman, D. (2011). Critical interventions in political toponymy. *ACME: An International E-Journal for Critical Geographies*, 10(1), 1–6.

Rose-Redwood, R., Alderman, D., Azaryahu, M. (2010). Geographies of toponymic inscription: New directions in critical place-name studies. *Progress in Human Geography*, 34(4), 453–470.

Rose-Redwood, R., Alderman, D., Azaryahu, M. (eds) (2018). *The Political Life of Urban Streetscapes*. Routledge, London.

Rose-Redwood, R., Vuolteenaho, J., Young, C., Light, D. (2019). Naming rights, place branding, and the tumultuous cultural landscapes of neoliberal urbanism. *Urban Geography*, 40(6), 747–761.

Sorokina, A. (2020). Pourquoi des stations du métro de Moscou ont-elles été renommées ? *Russia Beyond* [Online]. Available at: https://fr.rbth.com/histoire/85537-stations-metro-moscou-renommees.

Tomasik, P. (2018). Informative function in the names of urban transport routes. *International Scientific Conference Onomastic Investigations*, Riga, 120.

Woznicki, K. (2003). Saddam International Airport/Baghdad International Airport: The airport as a symbol and elementary unit of (neo)-imperial world land development [Online]. Available at: https://www.springerin.at/en/2003/2/saddam-international-airportbagdad-international-airport/.

Yaye Saidou, H. (2018). Toponymie populaire et transport à Niamey. Nommer les lieux en Afrique : enjeux sociaux, politiques et culturels. Pour un observatoire des néotoponymies urbaine, géopolitique et numérique. *Symposium International*, Niamey.

# 9

# The Toponymy of Informal Settlements in the Global South

**Melissa WANJIRU-MWITA**
*Technical University of Kenya, Nairobi, Kenya*

## 9.1. Introduction

Cities of the Global South are characterized by the presence of informal settlements. In some cases, these settlements hold up to half of the population living in those cities. Thus, they are an important part to take into account when considering the social, economic and cultural aspects of the city. While the term informal settlement is relatively well accepted, it is usually used as a synonym for slums. However, these two terms can be distinguished. Informality points to the insecurity of land tenure – lack of proper documents to ascertain outside of existing planning regulations and standards (Githira 2016). The idea of slums, on the contrary, highlights the dilapidated conditions of these areas and the low standard of living. Hence, slums qualify as informal settlements, but not all informal settlements are slums. Unfortunately, the use of the term informal has been applied indiscriminately, obscuring the diversity and complexity of these settlements. There is also a perpetuation of negative perspectives and interpretations of how these settlements operate (Lombard 2014). Current definitions tend to ignore the important aspects of the social organization of life in the settlements, hence falling short of a holistic understanding of these areas, whose unique characteristic is the

*The Politics of Place Naming*,

intertwinement of work and personal life (Chulek 2019). There is also a lack of visibility both on maps and in the urban landscape designs, leading to a deficient view of the settlements (Kamalipour and Dovey 2019).

Naming a place (toponymy) gives it some stability both spatially and symbolically. A name gives reference to a place in speech and exchange and separates it from other spaces (Leimdorfer et al. 2002). Toponymy since its origin is the meeting point between linguistics, geography and history. Place names describe spaces as they are or as they were, thereby bearing witness to different human activities, past or present (Dorier-Apprill and Van Den Avenne 2002). Place names portray a realistic picture of what happens in informal settlements and how the people living there relate to their immediate spaces. There lacks elaborate scholarship showing the importance of place naming in the informal settlements. Such research would serve to shed light on some of the contested images and expressions associated with the settlements. Many "informal" names have a negative connotation – reflecting the challenges faced by residents including poor living conditions, forced evictions and demolitions, crime and drug abuse, ethnic clashes inter alia. In informal settlements, the toponymic processes are mainly bottom-up in nature, that is, places are named by the residents based on the understanding of their environment in a local, organic and spontaneous fashion. This unique toponymic approach provides a theoretical imperative for studying place naming processes in informal settlements as a field within critical toponymy. Critical toponymy is a branch of toponymy that focuses on the socio-political processes of naming rather than the apolitical and encyclopedic study of names (Azaryahu 1996; Berg and Vuolteenaho 2009). The naming processes in informal settlements portray the Gramscian approach that emphasizes historical authenticity and equitable representation of all social groups. Gramsci sought to create an understanding of the processes of top-down coercion versus popular consent in explaining power dynamics in toponymic endeavors. He calls for organicity and reciprocity in place naming through incorporation of vernacular practices and recognition of local heroes (Vuolteenaho and Puzey 2018). The toponymy of informal settlements uniquely provides a window for assessing naming processes that occur from the "bottom" involving marginalized groups especially in urban areas. Designations and statements are mainly formed orally, with reference to social norms unique to those places. These popular modes of toponymic expression are seen to "contaminate" rather than "oppose" the initiatives of "graphic reason" (Bertrand 2001, p. 6).

The unprecedented growth of informal settlements has been caused by many factors including: population growth, rural–urban migration, wars and political unrests, ethnic clustering and lack of policies to deal with the exacerbation of these settlements (Jones 2017). The large population housed in these settlements is often referred to as "the unseen majority". They are unseen due to several factors including the lack of accurate population data, and the omission or partial representation of the settlements in topo-cadastral maps. This is partly because of reliance on traditional methods of mapping using information from census data or remote sensing (Mahabir et al. 2018). It is also a result of a lack of political goodwill to include these settlements in official maps or a willful omission so as to enhance the narrative that these settlements are illegitimate. In addition, these settlements are spontaneous and mushroom quickly. Whereas topo-cadastral maps are updated every few years, these settlements can start, grow and even be destroyed within a matter of months; thus, they are not always accurately represented in maps.

The persistence of informal settlements on the urban landscape has also has led to many remedial strategies being devised by governments and non-governmental organizations. These include incremental upgrading, infrastructural provision, and resettlement, and locally, to residents organizing legalization and upgrading including through community mapping (Choplin and Lovizit 2019). Other times, undesirable methods such as demolitions and evictions have been utilized leading to a vicious cycle of protests and skirmishes. Despite the many efforts to improve, upgrade, alleviate or eradicate informal settlements, they have remained an enduring part of the urban landscape in the Global South and thus need to be understood in their historical and cultural context.

## 9.2. Toponymy and informality – a theoretical background

Informal settlement struggles brought about by poverty, poor living conditions, constant evictions and demolitions have often been symbolically expressed through names. This way of place naming has been referred to as toponymic resistance or toponymic warfare (Kadmon 2004), whereby names are appropriated by marginalized groups to put up resistance campaigns against oppressive authorities (Rose-Redwood et al. 2010). This manifests more in informal settlements because the communities have vernacular geographical navigation and demarcation systems as well as informal jurisprudence and governance systems. Hence, they create other toponymic layers, below or beside the formal names. In some instances, formal names do not even exist in these settlements. The lower or alternate toponymic layers are more effective because they are used in daily life situations, bringing about a subtle toponymic resistance through use and lack of use of certain names. Myers, in his work on Zanzibar's Ng'ambo neighborhoods (*ng'ambo* literally translates to "other side" in Swahili), shows that these "other sides" of the city have "toponymic tapestries" and "micro-geographies" that indicate a unique spatial power dynamic that exists within the informal settlements (Myers 1996). The word tapestry points to the complex structure of the settlements made up of smaller geographical units. Informal settlements have largely been considered the "other sides" in cities of the Global South, separated from the main part of the city by physical barriers such as river valleys, highways or railways and socio-economic barriers of poverty, crime and bad living conditions.

Toponymic practices may take many forms, and involve different actors, including the state power, private sector and the civil society, that is, including the people themselves (Giraut and Houssay-Holzschuch 2016, p. 9). The spatial politics of informal settlements as reflected in toponymy show a resident produced approach of naming that emanates from the people and their local political systems. They tend to be influenced, on the one hand, by hardship and, on the other hand, by common unifying ideals. The people in slums are always in a state of alertness, driven by the prevailing discourses of conflict and contestation. There is a sense of temporality owing to the constant fear of demolitions or evictions. This reality calls for the residents to always be prepared for a change in occupation and residency (Simone 2004). Hence, the residents of these settlements are constantly fighting for legitimacy and belonging through various means including naming practices. Hence, their naming practices either reflect the actual situation, or aim to project a positive situation – perhaps a future desired scenario. Informal toponyms should not be set against formal toponyms because they serve a unique and important part of the urban population, namely, the urban poor. These names serve to reflect the everyday life of these residents and their relationship with space, thereby consolidating the understanding of collective heritage and contemporary challenges (Cumbe 2016).

## 9.3. Naming patterns in Nairobi's informal settlements

Informal settlements provide a unique geo-political context for the emerging toponymic scholarship in the Global South. The naming patterns provide a unique lens into a section of the urban sphere that is often overlooked or actively hidden and stigmatized. Nairobi provides a representative case study because approximately half of the total population live in informal settlements. The city is also ethnically diverse and has a unique position as a regional gateway and business hub for East

Africa. The toponymy of informal settlements is highly influenced by what occurs in the rest of the busy cosmopolitan city. Some of the toponymic categories that emerge from the informal settlement names include importation, formalization, ethnicity, commemoration, layering, environmental toponymy and socio-economic (live-work) relationship.

### 9.3.1. *Toponymic importation*

One important toponymic phenomenon is the importation of names from other places as a social reproduction of space through perceived notions of what is happening elsewhere in the world based on media reports. These local, partially informed naming practices exemplify how some counter-hegemonic struggles by subaltern and marginalized groups are enacted through naming across national and continental borders (Adebanwi 2012). Toponyms in this case are used as symbolic ammunition to fight against oppressive authorities, act as reminders of past injustices, and depict solidarity with other suffering people around the globe. For instance, the name Soweto is almost synonymous with the word informal settlements in Nairobi. Soweto is an acronym for South Western Townships, the largest urban Black townships in South Africa (Pirie 2016). There are at least three settlements in Nairobi named Soweto. The first Soweto in Nairobi is a sprawling slum area located in the eastern part of the city called Embakasi. The other two Sowetos (Soweto East and West) are located within Kibera; it is estimated to be the largest informal settlement in Kenya. Other interesting names found in Nairobi settlements are Kosovo, Vietnam, Gaza and Benghazi (Figure 9.3). In Cape Town, South Africa, there are two settlements named Beirut and Kosovo. All these countries, cities or regions have had globally renowned conflicts at some point in history, making the informal settlements' residents identify with them. Hence, they become toponyms of activism against governments that are largely perceived to be oppressive to the people in the informal settlements. The toponymic importation also depicts the use of place naming as a strategy to reflect the social and political struggles in the settlements, and establishes their toponymic imprint within a wider global phenomenon of resistance. Other international names are adopted to act as an inspiration for what people would like the place to become, highlighting the contrast between proper living conditions and what they currently experience. For example, in Cape Town, South Africa, there are two informal settlements called Europe and Barcelona (CORC 2016). They are named as such to indicate the kind of life the residents aspire to have now or in the future.

### 9.3.2. *Toponymic formalization*

Toponymic formalization occurs when names of informal places are adopted to represent more formal entities or formal administrative areas. As shown in Figure 9.1, the use of the name Soweto Slums has also led to the creation of other place names for more formal establishments. In this case, the figure shows the Soweto Slums, and in close vicinity is Soweto Catholic Church and Soweto Police Station. When previously unrecognized names are adopted by formal institutions, they acquire a level of legitimacy that allows them to slowly infiltrate the formal toponymic landscape. This leads to fluidity and overlap between the formal and informal toponymy of the city. Hence, toponymy should not be limited to physical locations, but has to be understood within a framework of institutions, community facilities, infrastructure, and the spaces in and around them.

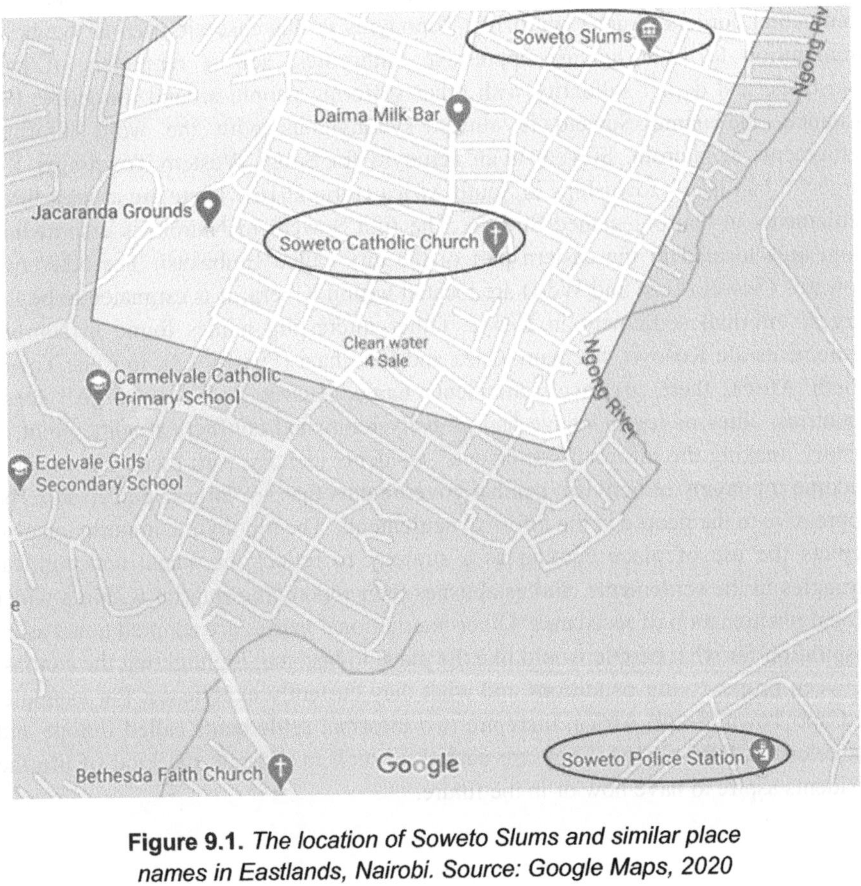


**Figure 9.1.** *The location of Soweto Slums and similar place names in Eastlands, Nairobi. Source: Google Maps, 2020*

### 9.3.3. *Toponymy and ethnicity*

Toponyms are seen as the meeting point of linguistics, geography and history. This is because place names are descriptive of the condition of places and the activities that have occurred there, past and present, and they also inscribe the nomination tendencies of different linguistic groups (Leimdorfer 2005). The micro-geographies of informal settlements provide a unique way of understanding the intricacies of urban ethnic conurbations. The settlements are highly cosmopolitan, with different ethnic communities cohabiting within very small spaces. Toponymy is one of the ways in which the linguistic diversity of these neighborhoods is reflected. According to the Nubian Council of Elders (NCE), Kibera was first occupied by the Nubian community who came from Sudan to work for the British government in the colonial period as part of the Kings African Rifles (KAR) during World War I. They were allocated the land on which Kibera stands today to set up their residences as camps in the early 1900s (Kibera NCE 2015) and gave their camps and villages Nubian names. Some of these names have persisted to date, others have changed in pronunciation, while others have been replaced with local names. Names such as Kibra, Makina, Lindi, and Kambi Muru are Nubian names that have remained. The name Kibera was formerly Kibra (meaning jungle in the Nubian language) and Laini Saba was previously Lain Shabaha (meaning rifle range area). These names were slightly distorted by local Kenyan communities who could not pronounce Nubian names properly. However, the name Kibra was restored and formalized when Kibra Constituency was created. While the Nubian heritage of Kibera still exists partly in the names of streets, villages and businesses (Figure 9.2), some of it has been erased, and the rest has been distorted resulting in a blended toponymy. Kisumu Ndogo (Small Kisumu) in Kibera is an example of imported ethnic toponymy. Kisumu is a town in western Kenya which is mainly occupied by the Luo community, one of the biggest ethnic groups in Kenya. The area occupied by Kisumu Ndogo was mainly inhabited by Luos hence the name (Kibera FGD 2016). In Mathare, another major informal settlement in Nairobi, there is a village referred to as Mandera. This is because it is mainly occupied by the Somali ethnic group, who originally come from the north-eastern part of Kenya where Mandera is located. The names portray ethnic clustering and also show a level of local toponymic importation, whereby the ethnic groups tend to name places after their places of origin; that is, Nubians gave their camps and villages Nubian names, Luos named their village "Small-Kisumu", and the Kenyan-Somalis named their area Mandera. Interestingly, names like Kisumu Ndogo, which was previously found in Kibera and other informal settlements, have now found their way to formal neighborhoods such as Donholm, removing the previous stigma of poverty associated with the name and giving it a level of formality. These names also highlight the rural–urban continuum and attachment of urban residents to their places of origin. This attachment is also reflected in the naming of businesses, for example, small shops and restaurants after the owners' rural villages or towns.

**Figure 9.2.** *A local restaurant with the name Nubian Cuisine in Kibera, Nairobi. Source: author's photo, 2015*

### 9.3.4. *Toponymic commemoration*

The commemorative aspect of toponymy aims at celebrating pioneers who are considered to be founders of a city, region or even village. Commemoration is one of the most commonly applied naming strategies both in formal and informal parts of cities in Africa and around the world. This is not without its challenges as some of the names can appear to be contextually or politically inappropriate. In highlighting some of the issues surrounding the commemorative naming of informal settlements in South Africa, Seabe states: "Our democracy is a conundrum of socio-economic ironies and injustices. Our poor stay in squatter camps (informal settlements for those who care to be politically correct) and our freedom fighters are honoured by getting squatter camps named after them." He goes on to add: "It is a slap in the face of the poor to name informal settlements after those who fought against an oppressive regime" (Seabe 2011). These sentiments seem to suggest that it is improper and ironical to name informal settlements after freedom fighters when the struggles have not yet been won. The South African case points to the controversial nature of commemoration of national leaders in informal marginalized areas of the city. Such naming can also be considered as a public address to political leaders, urging them to improve the conditions (e.g. settlements named after Joe Slovo, a former Minister for Housing in South Africa), or as a way to tactically shield those settlements from being destroyed; for example, demolishing a place named after Nelson Mandela would bring negative publicity for the authorities.

**Figure 9.3.** *The name Kosovo on a building in Mathare informal settlements. Source: author's photo, 2016*

Globally, the case of places that are named after Nelson Mandela is highly paradoxical. On the one hand, numerous large and well-known informal settlements are named after him in Brazil, Colombia and Sudan. This is done as a way for the marginalized communities that reside there to acquire some form of legitimacy and claim land rights. On the other hand, in the same metropolitan areas where the informal settlements are found, there are globalized formal areas in the central business districts which are named after the global icon and Nobel Laureate such as the Nelson Mandela Forum in Florence, Italy.

In the Kenyan context, the Mathare informal settlement has some of its villages named after previous local leaders who were deemed to have "discovered" the areas and pioneered their development. Some of the villages named after local leaders are KiaNdururu, Kwa Josphat and Kwa Nyang'au (Wanjiru and Matsubara 2017). The settlement also has a major road cutting through it which is known as Mau Mau Road, in remembrance of the Mau Mau movement that fought against the British colonial government in Kenya. These commemorative aspects of naming in the informal settlements are also a way of founding a toponymy that celebrates local (settlement founders and village chiefs) and national heroes (freedom fighters and elite politicians).

### 9.3.5. *Toponymic layering*

Informal settlements have a layered toponymy of macro-, micro- and alter-toponyms. In some sections, the place names reflect the social vices that are experienced. These micro-toponymies of smaller areas in the settlements develop spontaneously and reflect the changing physical and environmental conditions as well as the evolving ethnic composition to include people of other nationalities. For instance, a section of Mathare settlement is referred to as Kampala, the capital city of Uganda which borders Kenya to the west. On a superficial level, there is nothing to suggest that the name is associated with a negative or illegal business; however, according to Mathare residents, that part of the settlement has many immigrants from Uganda who are employed in the illegal liquor brewing business. Two other sections of the settlements are known as Nigeria and Cameroon because they are associated with illegal drug dealing businesses and have many illegal residents from the two countries (Mathare FGD 2016). Very often, the actual meaning of a toponym in an informal settlement is not immediately accessible to people who do not belong to that community. The naming acts as a way to control their space, that is, by keeping the real significance of a name to the insiders, and thereby creating an "us versus them" binary of mutual exclusivity. The effect of globalization and international migration in these settlements seem to have negative effects and as a result, as the names denote other geographical regions and countries, they also connote the negative aspects that residents associate with those places. Some areas within Mukuru settlement are called Jamaica and Kingston. This is because of the association the residents have given between the smoking of *bhang'* (marijuana) and their perception of that being a common phenomenon in Jamaica as depicted in the Rastafarian culture especially through reggae music. People coming from other countries are considered as outsiders, and the places where they occupy are named as such by the locals in order to demarcate those areas as being occupied by "others". In this way, naming acts as a tool for othering these incoming residents of the settlements. Superficially, the names of the international places may seem to indicate inclusion of those people groups, but a closer look shows that the names are used to mark those groups and create for them exclusive toponymic zones.

### 9.3.6. *Toponymy vis-à-vis economic and environmental conditions*

The Mukuru slum expressly portrays the "residence work syndrome" (Jain 2009) of many people in these settlements who are employed in nearby industries, since the settlement is located near Nairobi's industrial area (Mukeku 2018). As a result, some of the names in Mukuru such as Bins, Diamond and Gateway are named after companies in the industrial area. Work and home life are highly intertwined, causing the residents to name their villages after the industries where they work. The reason why people seek housing in Mukuru is because of the proximity to the industries, reducing transportation and housing costs. Unfortunately, some of the industries are responsible for pollution in the settlements near or within the living quarters.

The fragile environmental conditions in these settlements also have a big impact on the toponymic landscape. Most of the settlements are located on riparian reserves or flood prone zones, where dense housing construction increases the risk of disasters (Jones 2017). The word Mukuru itself means valley in one of the local dialects, already pointing to the nature of the location where the settlement is situated (Mukuru FGD 2016). In addition, the use of poor housing materials such as polythene paper and cartons, coupled with illegal and dangerous electricity connections, increase the risk of fire disasters. In Nairobi's Mukuru slums, for example, there is a village called Moto Moto (a Swahili word that implies calling out "Fire! Fire!"), a name acquired because of repeated fire occurrences in the area. Similarly, in Congo, Brazzaville, there is a settlement known as Poto-Poto meaning "watery mud". It was named as such because the neighborhood was located in a low lying, marshy area that was prone to flooding during the rainy seasons. Tsunami TRA/IDA (Temporary Relocation Area/Incremental Development Area) is the name of an informal settlement in the Western Cape region in South Africa. The name Tsunami is associated with the 2004 Indian Ocean earthquake and tsunami. In France, the Calais Jungle "Europe's Largest Slum" was a migrant settlement that developed between January 2015 and October 2016. It was mainly occupied by refugees and undocumented migrants awaiting to relocate to the UK or have their documents processed by the French government (Mould 2017, p. 388). The camp is located in a moorland area and was named in a derogatory way to describe it as being unplanned and unwanted. The term jungle has been used for other similar camps elsewhere in France.

## 9.4. Actors involved in the toponymy of informal settlements

Names are produced and used in different ways. There are several actors in the naming landscape of informal settlements. The first agents of toponymic inscription are the initial inhabitants of the settlements, also known as pioneers. This would often be politicians and village leaders who allocated land to the incoming residents. In other cases, it is people who contributed to the development of the settlements – such as educators or religious leaders. The other naming actors include the younger generation who are exposed to international media and know what is happening around the world. They adopt names of other places that have similar characteristics to the spaces they occupy. Government agencies and other formal institutions play a somewhat smaller role in the naming of informal settlements. This occurs when they establish schools, hospitals, community halls and other facilities that influence the names of the areas around them. Finally, the transport industry plays a unique role in influencing toponymy. Kibera, for example, is also known as *namba nane* (number eight) which is the route number for the buses that ply the Kibera area. Names of bus stops also tend to be used for surrounding areas, thereby creating a layer of names primarily influenced by the transport industry.

## 9.5. Conclusion – towards a toponymic framework for informal settlements

The case study analysis of informal settlements in the Global South in this chapter reveals new toponymic themes that are worth considering within the wider critical toponymic scholarship. These include toponymic warfare, importation, layering, and formalization. (1) Toponymic warfare points to the conflicting groups of the oppressed and the oppressor, and sometimes there are divisions among the oppressed groups. This leads to place names that express the conflict and divisions as well as the efforts of resistance. (2) Importation is in many cases a result of warfare. Most of the names portraying the resistance of the residents are imported from other international contexts especially those which underwent civil war. In this way, the residents show their understanding of global issues and their identification with them. (3) Toponymic layering has occurred where micro-geographies of unique socio-economic characteristics emerge within the informal settlements. These are given their own names that sometimes overlap with the existing toponymies causing multiple layers of names. (4) Toponymic formalization occurs when an informal name is used for a long period of time and eventually the name is adopted by the city authorities and used as a formal name to identify an administrative district.

Based on the existing toponymic literature, the naming patterns analyzed and the toponymic themes deduced in this chapter, we can see that toponyms act as symbols of oppression, ethnic delineators, indicators of resilience and markers of adaptation to socio-political, economic and environmental problems. In addition, unlike in the formal parts of the city, the residents of these settlements play the biggest role in naming. While this is often called a bottom-up approach, we can also describe it as a bottom for bottom approach because the people name their spaces primarily for their own consumption.

Naming practices in informal settlements bring to the fore novel theoretical concepts. Unfortunately, this scholarship has been neglected because most studies are based on the experiences of the Global North. Therefore, more attention should be given to the peculiar naming in informal settlements in the Global South as a new frontier in critical toponymic scholarship.

## 9.6. References

Adebanwi, W. (2012). Glocal naming and shaming: Toponymic (inter)national relations on Lagos and New York's streets. *African Affairs*, 111(445), 640–661.

Azaryahu, M. (1996). The power of commemorative street names. *Environment and Planning D: Society and Space*, 14(3), 311–330.

Berg, L. and Vuolteenaho, J. (eds) (2009). *Critical Toponymies: The Contested Politics of Place Naming*. Ashgate, Farnham.

Bertrand, M. (2001). Dynamiques urbaines, composition toponymique : le cas de Bamako (Mali). In *Nommer les nouveaux territoires urbains*, Electronic edition, Rivière d'Arc, H. (ed.). Éditions de la Maison des sciences de l'homme – Editions de l'Unesco – Open Editions, Paris.

Choplin, A. and Lozivit, M. (2019). Mettre un quartier sur la carte : cartographie participative et innovation numérique à Cotonou (Bénin). *Cybergeo: European Journal of Geography*, May.

Chulek, M. (2019). Hustling the mtaa way: The brainwork of the garbage business in Nairobi's slums. *African Studies Review*, 63, 1–22.

CORC (2016). Enumeration report – Barcelona informal settlement pocket [Online]. Available at: https://www.westerncape.gov.za/sites/www.westerncape.gov.za/files/hs-enumeration-study-wcg-ep-barcelona-final.pdf [Accessed on 8 July 2020].

CORC (2017). Enumeration report – Tsunami TRA/IDA [Online]. Available at: https://www.westerncape.gov.za/sites/www.westerncape.gov.za/files/hs-enumeration-study-wcg-ep-tsunami-final.pdf [Accessed on 8 July 2020].

Cumbe, C. (2016). Formal and informal toponymic inscriptions in maputo: Towards socio-linguistics and anthropology of street naming place names in Africa. In *Place Names in Africa*, Bigon, L. (ed.). Springer, Cham.

Dorier-Apprill, E. and Van den Avenne, C. (2002). Usages toponymiques et pratiques de l'espace urbain à Mopti (Mali) – La toponymie entre linguistique et géographie. *Marges linguistiques*, 3, 152–158.

Giraut, F., and Houssay-Holzschuch, M. (2016). Place naming as dispositif: Toward a theoretical framework. *Geopolitics*, 21(1), 1–21.

Githira, D. (2016). Growth and eviction of informal settlements in Nairobi. Master's Thesis, University of Twente, Enschede.

Jain, A. (2009). *Urban Housing and Slums.* Readworthy Publications (P) Ltd, Delhi.

Jones, P. (2017). Formalizing the informal: Understanding the position of informal settlements and slums in sustainable urbanization policies and strategies in Bandung, Indonesia. *Sustainability*, 9, 14–36.

Kadmon, N. (2004). Toponymy and geopolitics: The political use – and misuse – of geographical names. *The Cartographic Journal*, 41, 85–87.

Kamalipour, H. and Dovey, K. (2019). Mapping the visibility of informal settlements. *Habitat International*, 85, 63–75.

Kibera FGD (2016). *Issues Facing Kibera Residents*. Kibera Community Leaders, Kibera, Nairobi, 27 May.

Kibera NCE (2015). *Origin of Names in Kibera*. A focus group discussion held at the Kibera Mosque, Nairobi, 7 September.

Leimdorfer, F. (2005). Des villes, des mots, des discours. *Langage et société*, 114(4), 129–146.

Leimdorfer, F., Couret, D., Kouadio N'Guessan, J., Soumahoro, C., Terrier, C. (2002). *Nommer les quartiers d'Abidjan. Les divisions de la ville*, Topalov, C. (ed.). Éditions de la Maison des sciences de l'homme, Paris.

Lombard, M. (2014). Constructing places: Place-making in urban informal settlements in Mexico. *Progress in Planning*, 94, 1–53.

Mahabir, R., Agouris, P., Stefanidis, A., Croitoru, A., Crooks, A. (2018). Detecting and mapping slums using open data: A case study in Kenya. *International Journal of Digital Earth*, 13(6), 683–707.

Mathare, FGD. (2016). *Issues Facing Mathare Residents*. Mathare Community Members. Mathare Police Post, Nairobi, 2 June.

Mould, O. (2017). The Calais jungle: A slum of London's Making. *City*, 21(3–4), 388–404.

Mukeku, J. (2018). Urban slum morphologies and socio-economic analogies: A case study of Kibera Slum, Nairobi, Kenya. *Urbanisation*, 3(1), 17–32.

Mukuru FGD. (2016). *Issues Facing Mukuru Residents*. Mukuru Community Members. Mukuru, Nairobi, 24 May.

Myers, G. (1996). Naming and placing the other: Power and the urban landscape in Zanzibar. *Tijdschnft voorEconomisch a Sociale Geografie*, 87(3), 237–246.

Pirie, G. (2016). Letters, words, worlds: The naming of Soweto. In *Place Names in Africa*, Bigon, L. (ed.). Springer, Cham.

Rose-Redwood, R., Alderman, D., Azaryahu, M. (2010). Geographies of toponymic inscription: New directions in critical place-name studies. *Progress in Human Geography*, 34(4), 453–470.

Seabe, M. (2011). What's in the name of informal settlements? [Online]. Available at: https://www.dailymaverick.co.za/opinionista/2011-10-15-whats-in-the-name-of-informal-settlements/ [Accessed on 4 February 2020].

Simone, A. (2004). People as infrastructure: Intersecting fragments in johannesburg. *Public Culture*, 16(3), 407–429.

Vuolteenaho, J. and Puzey, G. (2018). Armed with an encyclopedia and an axe: The socialist and post-socialist street toponymy of East Berlin revisited through Gramsci. In *The Political Life of Urban Streetscapes: Naming, Politics and Place*, Rose-Redwood, R., Alderman, D., Azaryahu, M. (eds). Routledge, Abingdon.

Wanjiru, M. and Matsubara, K. (2017). Slum toponymy in Nairobi, Kenya: A case of Kibera, Mukuru and Mathare. *Urban and Regional Planning Review*, 4, 21–44.

# 10

# The Map, the Name and the Territory: Toponymic Struggles in the Era of Cartographic Post-Sovereignty

**Matthieu NOUCHER**
*CNRS Passages, Bordeaux Montaigne University, France*

## 10.1. Place names, an issue of information sovereignty

In the official letter sent on February 5, 2018 to MP Valéria Faure-Muntian, French Prime Minister Édouard Philippe explains the motivations for the mission he has entrusted to her: “Spatial data remains more than ever a vector for informed action by the State and a sovereignty issue.” The expression “sovereign data” subsequently comes up six times in this three-page letter which details the expectations for this study on the governance of spatial data in France. This would lead, a few months later, to the submission of a report entitled “Sovereign Spatial Data” (Faure-Muntain 2018). This question is of concern to public authorities around the world today as the digital revolution profoundly reconfigures the strategic environment of states and the way in which they attempt to control the “datasphere” (Douzet 2020). Mapping therefore appears particularly symptomatic of these new digital governance issues. In fact, the field has experienced a series of technical, legal and organizational reconfigurations over the past 15 years that have overturned the way in which maps are produced, distributed and used. The digital turn and then the development of the Internet with the emergence of the geoweb – defined, at least, as the aggregate of spatial technologies and georeferenced information

*The Politics of Place Naming*,
coordinated by Frédéric GIRAUT and Myriam HOUSSAY-HOLZSCHUCH. 

organized and transmitted via the Internet – have given rise to new actors, new practices and new tools. As a result, mapping is no longer the sole responsibility of States. Multinationals like Google as well as communities arising from digital commons like OpenStreetMap have become key players. I therefore argue that, despite the State's very voluntarist rhetoric seeking to rehabilitate a form of cartographic authority, the recent evolution of the geographical information landscape leads to the emergence of a form of cartographic post-sovereignty.

Cartographic sovereignty can be defined as the exclusive privilege of states to provide their own representation of the national territory. This is reflected in the divergent technological choices of national cartography, even in the geodesic systems or the units used. The state thus constitutes itself as an essential source of cartographic authority through its ability to control the collection, representation and also the distribution of official maps. However, today's geoweb causes a shake-up of sovereign digital geographies by multiplying potential sources as well as channels of distribution. Official maps are supplemented, circumvented or even competed with by alternative maps, and cartographic authority is called into question. The methods of production and circulation of place names are exemplary processes from this point of view. If official toponymy is a prerogative of local or national authorities, toponymies of use or even counter-toponymies emerge on the Internet to the point of completely blurring the maps, as I will explain in the first part.

To illustrate these issues, I propose to rely on the heuristic case of French Guiana. It presents, from the point of view of place names, a complex situation that is particularly interesting to analyze: settler place names from the time of the colonial conquest are juxtaposed with native place names. However, official maps far from reflect this diversity. They seem to be fixed in received ideas that have "stuck" to Guiana for centuries, while fostering them very widely (from *terra nullius* to green hell, from the El Dorado of the miners to the hostile jungle). Therefore, the geoweb appears as an alternative distribution channel where issues of knowledge and recognition of the different communities which make up French Guiana are expressed. By returning to some historical and recent cartographic controversies, I will highlight the permanence of the political stakes of the naming of this territorial margin. I will then show how the geoweb can open up new perspectives of resistance.

Finally, I propose a research agenda. I then defend the thesis that in addition to critical analyses of the meaning of *place names* or *place naming*, it is now also necessary to deconstruct the power games associated with the regulation and control of the circulation of the various existing toponymic databases (*place name flow*).

## 10.2. Cartographic post-sovereignty and place names: when the geoweb blurs the map

### 10.2.1. *Cartography and sovereignty: the double challenge of a critical approach*

Modern cartography, understood as large-scale topographic cartography, was constructed at the beginning of the 19th century, in a close relationship with state power (Desbois 2015a). Colonial cartography is a good illustration of this link between the affirmation of modern states and the representation of space, between territorial domination and control of the content of maps. The actors of colonization, in fact, participated very widely in the laundering of maps by erasing certain markers so as to create from scratch an empty space, which then became a space to be conquered (Laboulais-Lesage 2004). The colonial map therefore gives shape to space by marking it with elements (roads, waterways, etc.) that will allow it to be tamed. At the same time, it expels indigenous practices and memories into the void of a blank space. Place names are at the heart of these selective practices. Thus, Hélène Blais, in *Mirages de cartes*, analyzes, between bricolage and adjustment, the cartographic constructions at the origin of "the invention of colonial Algeria" (Blais 2004). Jordan Branch (2014), meanwhile, retraces the history of modern cartography in relation to the evolution of international relations. Drawing notably on the analysis of peace treaties, he exposes how the developments of maps and political sovereignty are crucially linked. Branch's thesis is twofold. On the one hand, he defends the idea that the representation of delimited and named spaces on the first modern maps preceded modern political practices based on these limits and on these names. On the other hand, it explains how the relationship between cartography and the development of modern territories shapes our understanding of international politics today. In *The Sovereign Map* (2005), Christian Jacob emphasizes this last point by examining the role of the *spectator* in decoding the meaning of a map and the role of society in defining the power of maps as authoritative representations of space.

On the strength of these observations, maps have for several decades been the subject of critical analyses that decipher from a political perspective their construction and their socio-spatial effects. These analyses are part of the current critical cartography, a movement that questions the map as a neutral and objective representation of reality, to conceive of it as a social product essentially used for purposes of domination. By distancing himself from functionalist approaches and privileging their cognitive scope, the Englishman John Brian Harley, in his reference text *Deconstructing the Map* published in 1989, invites us to rethink maps as forms of socially constructed, subjective and ideological knowledge. A historicist current would then emerge, in the years 1980–1990, in particular around the study of the map as an instrument in the service of the formation of the modern state. Other research has progressively nuanced the vision of cartography summed up solely in the "repressive" exercise of power, in favor of a Foucauldian approach for which the map becomes part of a "knowledge-power". More subtle forms of domination and control, circulating, internalized by individuals and no longer imposed by force, are then revealed. The role of cartography as one of the techniques of territorialization of power, as well as a vector of cultural and social mutation of the population, is then made clear. Once again, place naming, through its ability to construct and disseminate hegemonic visions by freezing certain denominations and masking others, is one of the key elements analyzed by critical cartographers.

Faced with the development of digital geography, these critical approaches find a new resonance with the communities of researchers who work on geographic information systems (GIS), develop geomatics and gather in the United States under the name of *GIScience*. The emergence of a current of research around *Critical GIS* is then marked by the rise of questions about the social implications of these technologies. Towards the end of the 1990s, critical GIS became a sub-discipline of digital geography (Desbois 2015b). Researchers defend the possibility of a critical use of GIS, in particular embodied in a feminist geography (Kwan 2002) or by the introduction of GIScience techniques into actions of participatory democracy with the implementation of participatory GIS (Rambaldi 2005). The issues of cartographic sovereignty, at one time removed from scientific concerns and the media scene, reappeared at the turn of the 2010s, with the development of the geoweb and the politicization of "digital sovereignty".

### 10.2.2. *The questioning of state authority or the emergence of a double cartographic deregulation*

The notion of "digital sovereignty" has been growing in interest since 2013 and Edward Snowden's revelations about the mass surveillance conducted by the United States. In response to these revelations and faced with the unchallenged domination of the Sino-American duopoly over the digital world, many European states have called for guaranteeing Europe's technological sovereignty. The question of data, from their collection to the control of their dissemination, including their processing, has become a key strategic issue. Cross-border platforms, whose added value is based on their capacity for information intermediation[1], are becoming prominent on the international scene and are gradually becoming an additional challenge to the

1. Intermediation platforms connect actors to allow them to exchange goods or services on two-sided markets, such as drivers and passengers in the case of Uber. Many platforms place informational content at the heart of the services offered.

digital sovereignty of states. These large digital platforms indeed shape the informational content that circulates on the Internet by selecting data, prioritizing results, directing flows, and opening or closing the circulation of data (Grumbach 2020). The informational space and control of the data circulating on the Internet are therefore no longer under the sole control of states (Douzet and Desforges 2018).

Cartographic data perfectly illustrate these recent developments and the challenges of the informational governance of territories. The development of GIScience, the growing availability of spatial data (in particular satellite imagery), the growing involvement of digital multinationals and then the meteoric rise of cartography on the Internet have considerably transformed the links between state sovereignty and cartographic production. Also, despite the state's voluntarist rhetoric to rehabilitate a form of cartographic authority – as evidenced, in France, by the recent parliamentary report quoted in the introduction – the evolution of the landscape of geographical information has led over the last 10 years to the emergence of a form of cartographic post-sovereignty. The sovereignty of states, which was exercised particularly through control of the data, which their national mapping institutes disclosed about their territories, is being disrupted by digital platforms with a global reach, but whose content is based upon very reactive local networks. This development of spatial data, in all directions, may appear both as the symptom and as the agent of the crisis of nation-states confronted with globalization (Desbois 2015a). A double deregulation is thus observed: from above when the state is in competition with planetary platforms and from below when it is circumvented by local actors.

#### 10.2.2.1. *Cartographic deregulation from above: states facing the giants of the (geo)web*

The convergence of GIS with information and communication technologies (Internet, mobile telephony) and geolocation systems (GPS) have led to the emergence of new methods of manufacturing and disseminating geographic information, which is reflected well in the notion of geoweb. The *geospatial web* or geoweb can be defined as the generalization of organization by the information space on the Internet. This development was rapid and unexpected because it did not come from institutional and technical players in the GIScience sector. Indeed, it is the massive investment of digital (Apple, Microsoft) and web (Amazon, Google, Facebook) multinationals in geospatial technologies that has led, since the mid-2000s, to a major reconfiguration of production and visualization of digital maps. The emergence of new cartographic interfaces, such as virtual globes, has met with an enthusiastic response both from the public and from the creators of Internet sites and mobile services. The ease of reuse of tools has led to the rapid dissemination of a few planetary platforms that have quickly become essential, such as Google Maps or Microsoft's Bing.

Beyond the new visualization interfaces, the Web giants have also invested in the production of spatial data by creating their own cartographic repositories. Deploying impressive collection infrastructures (from *Google Cars* surveying road networks while photographing landscapes, to satellite constellation projects), these multinationals do not simply manage the *pipes* through which information circulates; they also control the content circulating there. This control operates in two distinct ways: on the one hand, by producing their own geographic database, they supplant the cartographic reference systems of states by distributing planetary base maps. The "Google Maps style" thus leads to a homogenization of cartography which masks national graphic specificities (Lieselot et al. 2020). On the other hand, to supplement their own data, these platforms aggregate multiple sources. In doing so, they select and prioritize information. Once again, place naming appears as an exemplary case of these new methods of data control. Thus, as Jack Nicas notes in the *New York Times*[2], in San Francisco, a place branding operation led shopping centers to be renamed on Google Maps, without any consultation with local authorities, due to certain sectors of the city having a bad reputation. This virtual street naming quickly spread to the digital world (from hotel booking centers to dating sites, from the Uber delivery platform to real estate agencies) and then to the physical world with the appearance of advertising billboards. These surreptitious naming changes illustrate how Google has become the primary arbiter of place names. Decisions taken by a few cartographers from the Mountain View firm can reshape the identity of a city, a village or a district, without validation by the public authorities. This observation at the local level also applies at the national or international level. The names of countries and seas, like the drawing of borders, are regularly the subject of controversy. In defiance of international arbitration, Google in fact adapts the outline of borders and the names of countries/regions according to the location of the user consulting its maps.

This deregulation from above, which gives to a few rare players who dominate the geoweb an unprecedented ability to circumvent the cartography of states by imposing their own vision of the world, is coupled with a deregulation from below, less spectacular but just as effective.

2. August 2, 2018 edition.

#### 10.2.2.2. *Cartographic deregulation from below: the State faced with volunteered geographical information*

Investment by digital giants has led to an unprecedented boom in geospatial technologies and has also contributed to a form of democratization of cartographic tools: GIS reserved for specialists now cohabits with online cartography, virtual globes and other geoportals whose handling is greatly facilitated (Joliveau et al. 2013). To translate the technical and organizational reorganizations which shook up the world of GIScience under the influence of Web 2.0, Michael Goodchild published, in 2007, a reference text in which he proposed the expression "volunteered geographical information". This notion has gradually imposed itself as a unifying framework which reflects the central place now occupied by the citizen in these new modes of mapping.

Each Internet user can now consult, annotate or cross-reference their data by georeferencing them and mobilizing them on online cartographic interfaces. A profusion of ego-cartographies can thus be observed on the Internet with the dissemination of memorial and personal maps, or even with the visualization of digital traces (Noucher 2019). While solicited uses have appeared in the field of geomarketing, social networks, participatory sciences, crisis management or even citizen consultation, other more spontaneous practices have gradually emerged in the cultural, artistic or even political fields. Collaborative mapping projects like OpenStreetMap (OSM) now occupy an important place in the geographic information landscape. Often presented as the "Wikipedia of cartography", OSM is a project that offers thousands of contributors multiple collaborative environments to expand a single global cartographic database, freely editable and shared.

This ease of production and dissemination of spatial data on the Web also contributes to strengthening the capacities of non-experts in mapping to reclaim the power to map their own territory. Previously exclusively *top-down*, and thus an attribute of power, the production of maps is also gradually becoming a *bottom-up* production. Multiple actors, traditionally excluded from the cartographic production process – neighborhood residents, indigenous peoples, environmental movements, etc. – have thus appropriated the tools and modes of representation of cartography to put pressure on public policy. Though, as an extension of participatory mapping, participatory GIS have made it possible to introduce GIScience techniques into activities of participatory democracy since the 1990s, the rise of the Web tends to amplify this type of initiative (Joliveau et al. 2013). From Cotonou to Benin (Choplin and Lozivit 2019) or Saint-Laurent du Maroni in French Guiana (Morrachini and Comptour 2020), there are many examples of participatory mapping that mobilizes OSM to render visible neighborhoods (informal, marginal, spontaneous, etc.) that do not appear on "official" maps. Toponymic questions are therefore at the center of these projects which, through collaboration with the inhabitants of these districts, make it possible to identify the denominations in use. These latter may then testify both to historical episodes or anecdotes and to the linguistic and cultural diversity of places (Quinn 2016). The toponymies of inhabitants inscribed on the map also make it possible to give visibility to subaltern communities who live in these cartographic no man's lands.

The distinction between deregulation from below and deregulation from above must be nuanced, as interweaving between different registers of production is sometimes required. Thus, when Facebook announced the purchase, in June 2020, of Mapillary – a geolocated photo-sharing system based on the Google Street View model – it further confused the issue by becoming the private owner of what, for many contributors, nevertheless constituted a digital commons. A mapping war including toponymic issues is therefore in process today, with configurations that may be multiple but retain one constant: public power is systematically excluded. The political stakes of place naming must be analyzed in the light of this transition, muffled and brutal, towards cartographic post-sovereignty.

### 10.2.3. *(Re-)thinking the political issues of place naming in the era of cartographic post-sovereignty*

Since the geoweb has begun to blur the maps, how to understand place naming today? How to approach the naming of places when volunteered geographical information has transformed the figure of the cartographer into an elusive hydra? The long focus of critical cartography on the object of the "map" makes it difficult to understand these complex evolutions. The expansion of the uses of spatial representations well beyond the sphere of cartographers and other GIS specialists has not been accompanied by a renewal of this field of research; quite the contrary. The lack of involvement of geographers in tools for the "general public", of the virtual globe type, whose "technical" operating logic escapes them, has generated a form of disinterest. Moreover, following Pierre Gautreau (2021), we can consider that critical cartography has paid excessive attention to globalizing histories of major cartographic projects. In many respects, this immoderate taste for the epic still infuses most of the work within this movement. In his work published in 2004, Jean Lefort thus presents the history of the concepts and mathematical tools of cartography as "a human adventure, that of scholars and explorers, from their dreams to the adventures of their travels, from their vision of the world to the development of their instrument". Though these grand cartographic narratives have constructed the history of the discipline, they should not be a unique key to reading it today. This is why, in 2017, I proposed the expression "small web maps" (Noucher 2017). It reflects two main perspectives of evolution: historical and political. From a historical point of view, the idea of small maps makes it possible to mark the break between grand cartographic narratives, which have always guided the discipline, and the recent emergence on the Internet of diffuse and exponential practices. Moreover, the politicization of what is "small" also leads to considering "small web maps" as minor maps, which is to say, produced outside the dominant institutional or commercial spheres. They then potentially give voice to various minorities and may even become counter-maps. According to methods common to Subaltern Studies, it is then a question of opening the study of maps onto corpora that are currently ignored, and of analyzing the way in which these data circulate, in which they are appropriated or made invisible, in which control over their transmission reveals (new?) power games.

It is in line with this proposal that I suggest addressing the toponymic databases that permeate the geoweb and gradually penetrate the physical world. To better account for the political stakes of place naming in the era of cartographic post-sovereignty, I defend the idea that it is necessary to complete critical analyses on the meaning of *place names* or on *place naming* by studies, making it possible to account for the fine mechanics of their circulation (*place name flow*). Far from being anecdotal, I hypothesize that the multiple socio-technical mediations operated by institutional, commercial or community platforms that regulate, control, block or impose place name flow deserve to be analyzed. They potentially reveal choices that are more political than technical, as evidenced by the permanence of the issues raised by the study of place names in Guiana.

## 10.3. Toponymic struggles of yesterday and today: the exemplary case of Guiana

### 10.3.1. *Place names in French Guiana, a legacy of myths and conflicts*

As with cartography, place naming constitutes a "colonial technology" (Walker et al. 2013, p. 7) which has been at the heart of the conquest of the interior of Guiana since France set foot there in 1643 and then, definitively, in 1664. From the island of Cayenne, the French colonists progressed little by little along the coasts. The French colony therefore appears, at first, as essentially coastal. The forest represents a natural barrier, an implicit limit both on the ground and on the maps. The cartography of this space is nevertheless not empty: the blanks on 17th-century maps are thus filled by many imaginary figures[3] and by place names. Lacking precise locations, the toponymic filling-in of the map thus makes it possible to occupy the space, to mark a possession or to erase any previous presence. A number of maps[4] reveal, in this interior of Guiana left to the free interpretation of cartographers, archetypal visions of the colonial space (Regourd 2004). The case of the legendary Tumuc-Humac Mountains and the Franco-Brazilian disputed boundary offers two exemplary situations of the use of place names as vectors for the staging – in text and images – of colonization.

#### 10.3.1.1. *The myth of the Tumuc-Humac Mountains or when place names are used to "close" space*

Appearing from the beginning of the 19th century on Spanish maps, the Tumuc-Humac Mountains owe their popularity, in France, to the explorations of the military doctor Jules Crevaux (1847–1882) for whom they constitute an inaccessible ideal, a symbol of his trip to Guiana; then, to the stories of the geographer Henri Coudreau, who would go so far as to claim to have crossed the "high passes of the Tumuc-Humac" and draw up a map of them in 1887 (Figure 10.1). The geographer Jean-Marcel Hurault, son of the founder of the National Geographic Institute (IGN), would be the first to affirm that this map constituted a fake, drawn up without respecting distances and altitudes on routes that Coudreau had not actually traveled. From the 1950s, aerial photographs revealed the non-existence of a real mountain range. However, this place name, very practical from a geopolitical point of view for marking a *natural* border with Brazil, would appear on IGN maps until the end of the 1960s!

This type of cartographic myth, maintained by the public authorities, is not anecdotal in a region like the Guianas Shield which, even today, is the scene of

3. We might think here of the lions, headless humanoids and Amazons that border the imaginary Lac Parimé on the 1598 map of the Flemish engraver, cartographer (and artist!) Jocodus Hondius. The precision, for the time, of the description (geometrical and toponymic) of the coasts contrasts with the blanks of the map of the interior, which are then filled in by these legendary figures.

4. We might cite here, without seeking to be exhaustive, the printed map of Guiana by Du Val (1654), the map of Guiana by Lefebvre de la Barre (1665) or the printed map of Guiana by Bellin (1763), copies of which can be found in the National Library of France.

numerous land and sea border conflicts (de Vilhena Silva 2017). These conflicts are often settled with maps and counter-maps whose place names then constitute key elements of the arguments of the various stakeholders.

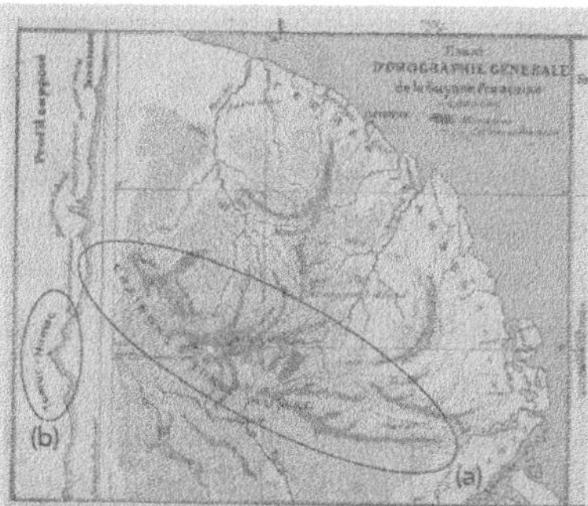


1887: Extract from "Map of Guiana. Exploration of the Tumuc-Humac, tributaries of the Maroni and the Oyapock, etc."

Published in the mission report for the Ministry of Public Instruction by Henri Coudreau.

This inset dedicated to orography allows us to note the precision of the outline of the links in the Tumuc-Humac chain, which appear as much on map (a) as on the North-South "presumed outline" (b).

Source: BNF – http://gallica.bnf.fr/ark:/12148/btv1b53027786w

1895: Engraving of the Tumuc-Humacs from Mount Ariqinamaye

Published by Hachette in the account of Henri Coudreau's voyage, this engraving was accompanied by the following commentary:

"Before me is a mountainous massif 300 kilometres long and 100 wide, as large as Belgium. In the three hundred years that we have possessed Guiana, we voyagers have not yet been able to report anything positive about this mysterious Tumuc-Humac chain. This massif must be discovered in its whole, studied in its details. To guide me, no written document, no barely accurate indigenous information, nothing." (p. 92)

Source: BNF – http://catalogue.bnf.fr/ark:/12148/cb34139685r

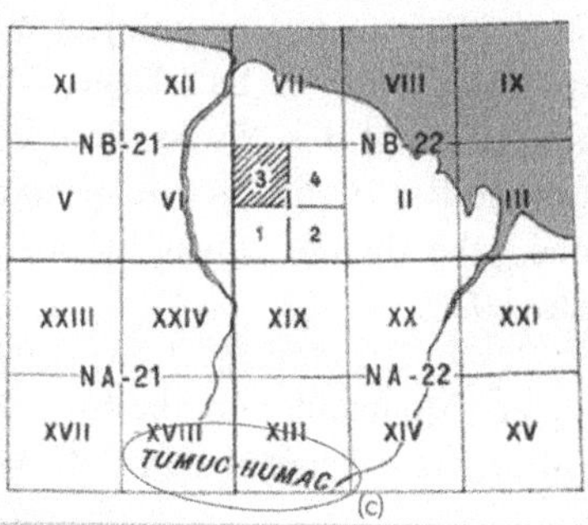


1950: extract from IGN's photogrammetric sketch

Created based on aerial photography by the US Air Force (1945), the IGN published a photogrammetric sketch at 1/100,000 scale in colour in 1950.

Tumuc-Humac is the only place name (c) used in the index opposite. It marks the southern frontier of Guiana. While the region is delimited by the Atlantic Ocean in the north and the Maroni (in the west) and Oyapock (in the east) rivers, the place name (c) that links their two sources seems to "close" the territory by "naturally" marking the southern frontier.

Source: Document library http://regards.cnrs.fr

**Figure 10.1.** *Appearance and persistence of the place name Tumuc-Humac*

#### 10.3.1.2. *The Franco-Brazilian disputed zone, or when place names are at the heart of a border conflict*

The delimitation of the border between France and Brazil gave rise to a veritable "battle of maps" where place names played a key role. Throughout the 19th century, the French and the Brazilians claimed sovereignty over an expanse of 200,000 $km^2$ of savannas and forests located between the rivers Oyapock, considered to be the border limit by Brazil, and Araguari, considered to be the border limit by France. Taking advantage of the official absence of political authority, many adventurers then settled there, as well as runaway Brazilian slaves; subsequently, from 1894, wildcat gold miners came from the Guianas and the West Indies following the discovery of a significant gold deposit in the region (Granger 2011). In 1895, a clash in the disputed zone between a French military detachment and a Brazilian militia caused about 40 deaths, mostly civilians, in the small town of Mapá (Rouard de Card 1897). Faced with this international scandal, France accepted the Brazilian proposal for Swiss arbitration to settle this border dispute definitively. In Bern, a French delegation that was little aware of the problem but passed on the maps and arguments drawn up by the geographer Vidal de La Blache faced off against Brazilians led by the diplomat Rio Branco, armed with an impressive corpus of historical maps (Granger 2011). Rio Branco used place names as proof not only of the location of the disputed river, but also of the antiquity and prior history of the Brazilian settlement, whereas Vidal de La Blache, with a more traditional use of maps, stuck to geomorphological arguments alone (Mercier 2009). The toponymic argument prevailed, and, in December 1900, the Swiss president awarded Brazil almost all of the disputed territory. The Treaty of Bern thus marked the triumph of Rio Branco's conception of geography, integrating place names as political and historical elements, over the more classic conception of Vidal de La Blache, who wanted to stick strictly to natural arguments. If few traces of this "Swiss sentence" are perceptible on the French side, in Brazil, Baron de Rio Branco is celebrated as illustrated by the street naming of Oiapoque, the border town where a commemorative monument also proudly sits (Figure 10.2).

The examples of the mythical Tumuc-Humac Mountains or the role of denominations in the arbitration of the Franco-Brazilian disputed zone are not anecdotal: they testify to the evocative power of place names, over historical episodes which have shaped and continue to forge imaginaries associated with Guiana. These imaginations today continue to feed into received ideas about a supposedly unstable region and will play an active role in the direction of its toponymic renewal.

**Figure 10.2.** *Tributes to Baron de Rio Blanco at Oiapoque, Brazil [photo credits: M. Noucher, November 2019]. Top: monument to the glory of Baron de Rio Branco and the Treaty of Utrecht in the streets of Oiapoque opposite the headquarters of the federal police. The shape of the sculpture in the upper part of the monument mimics the outline of the borders and under the medallion of Rio Branco, we can read in French and Portuguese the words of the Treaty of Utrecht of 1900. Bottom: street sign of the Avenue of Baron de Rio Branco in the streets of Oiapoque*

### 10.3.2. *Toponymic renewal: the State grappling with its received ideas*

Although cartographic myths and border conflicts still have a major impact on the toponymic landscape of French Guiana today, it is not completely fixed. Changes are thus regularly suggested to the IGN, which produces reference maps in France, and to the National Commission for Place Names, which oversees the conservation as well as the coherent development of the country's toponymic heritage. Holder of the prerogative of official names on the territory, the state therefore plays, through these bodies, a key role in attempts at toponymic renewal. This semi-regal competence operates remotely from Paris, in a territory, Guiana, still marked today by a number of received ideas.

#### 10.3.2.1. *Inner Guiana by its official place names: between terra nullius and green hell*

In a book published in 2016, the geographer Roger Brunet proposed a dictionary of French place names. Over more than 700 pages, it describes in detail the names of places in mainland France and overseas. Only one small page is devoted to French Guiana and the interior is quickly swept away with this sentence, without appeal: "Vast territories are populated neither by inhabitants nor by place names, at least on detailed maps." Brunet takes the wise precaution of specifying that his observation is valid "at least on detailed maps" because, as he explains in the preamble to his book, his method for collecting place names consisted of exploring the IGN Geoportal, the cartographic platform of the French State. However, by browsing the topographic map (1/25,000) of French Guiana on the Geoportal, the toponymic identification exercise must have been relatively quick…

Presented as the "French-style Google Maps" in a column signed in 2009 in *Les Échos*, by the director general of the IGN at the time, the Geoportal aimed to demonstrate that "the domination of the GAFA companies over geolocated information is not inevitable" and to defend "national sovereignty" through a public service of spatial data produced and hosted in France. Defending his institute vigorously, he emphasizes: "The State must have reliable data, supplied independently, to enable it to make decisions in a sovereign manner." It is in particular the high frequency of updates and the general and homogeneous coverage of the whole of the territory which should make it possible to distinguish the cartographic reference systems of the public service from private offers which, we sense, are only interested in regions with strong commercial potential. But, in browsing the topographic map of French Guiana, we can question the application of such a principle beyond mainland France, in that outside the coast, Guiana is dominated by blank space (Figure 10.3).

To explore inland French Guiana, Roger Brunet therefore had to change the scale of analysis and settle for the 1:500,000 map produced by IGN in 1995. As Lézy (2000) pointed out, the low diversity and poverty of place names on this map seems to freeze Guiana at the time of the conquest and the pioneers of wildcat gold mining. Place names are focused on gold sectors with placer[5] appellations with evocative names such as: "failure", "good hope", "patience", "certitude", "misery", "madness", etc. These denominations feed, in the minds of Europeans, the idea of a green hell. These symbols also continue to shape the perspective of a *terra nullius* and testify to the inability of these settler experiences to decode indigenous markers. However, if we look at the names used by the inhabitants, the current toponymic

5. A placer is a mineral deposit.

landscape of Guiana is characterized by its richness and diversity resulting from influences that are indigenous, Brazilian, Surinamese, Creole, Bushinengue, Hmong. The recent renewal of based maps, however, far from reflects this diversity.

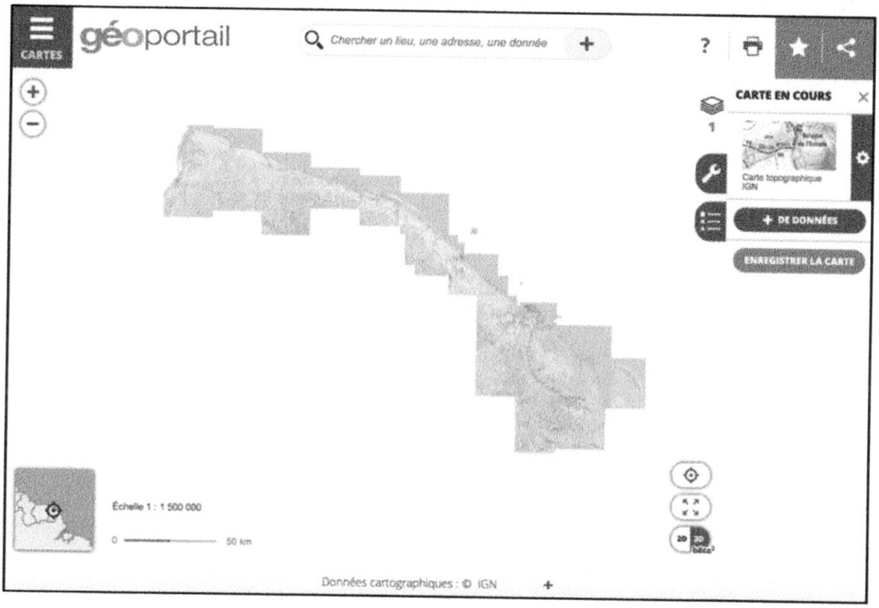


**Figure 10.3.** *Blank on the map*[6] *[source: https://www.geoportail.gouv.fr/ (accessed August 2020)]*

### 10.3.2.2. *When the military generate place names… without knowing*

In 2009, IGN with the support of the Ministry of Defence launched production, over the 84,000 km$^2$ of the territory, of a campaign of aerial photographs and a new cartographic database from which a series of maps at 1:50,000 were derived. At the end of 2011, the entire series was produced. This short production time is a point of pride for the Institute which recognizes, all the same, in its internal newsletter: “It is a relatively ‘expedited’ cartography that uses a maximum of automatisms and a

6. The display on the national geoportal of IGN topographic maps gives a unique image of the territory: with no neighbors and no connection with the ocean to the north or the forest to the south, French Guiana boils down to a coastal strip whose strange division extends to the west along the Maroni to Saint-Laurent, while to the east it even ignores the Oyapock, its border river. If this island-map is only a reflection of the weakness of the coverage of national topographical references, we may wonder if this autarkic vision does not contribute to fuelling the idea of an internal no man’s land.

minimum of interactive repetitions" (Loil 2011, p. 11). This expeditious nature is notable when traveling through the topography represented on the map: it is thus not uncommon to observe watercourses doing "roller coasters" and going up certain slopes (Noucher 2020a)!

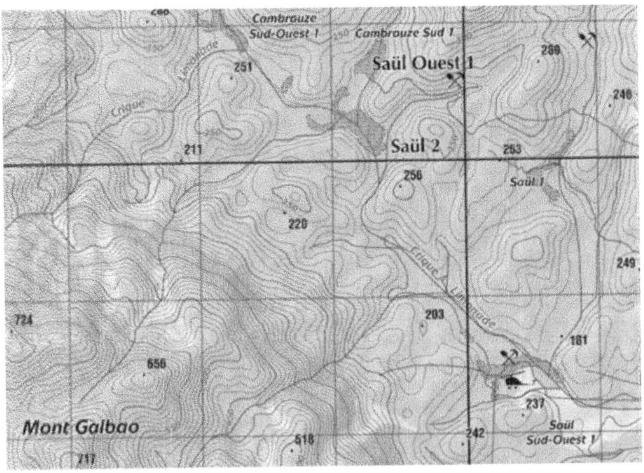


**Figure 10.4.** *Topographic map of the surroundings of Saül, French Guiana*[7] *(source: https://www.geoportail.gouv.fr/ (consulted in April 2020))*

The place names are also very approximate. Military sources have been preferred. Certain denominations, used by the Armed Forces to identify illegal wildcat gold mining sites, such as "Guerrilla", come from listening to radio from *garimpeiros*[8]: clandestine gold miners thus become the sources of official place names, despite themselves! In addition, a good proportion of the place names appears to be totally fanciful. For example, around Saül, there are sites that correspond to points geolocated by the French Guiana Armed Forces (FAG) and indexed as Saül_1, Saül_2, etc. (Figure 10.4). In fact, the FAG provided IGN with their entire database, including a series of geolocated points. These include in particular sites for landing a helicopter. IGN has integrated all of this data en masse. Today, the military are the first to be surprised to see these sites appear on the map and to note that the indices they quickly filled in have become place names[9]. The

7. Near Limonade creek, "indexed" place names Saül_1, Saül_2, Saül West_1, Saül South-West_1 from the database of the Armed Forces of French Guiana, deemed legitimate for publication on the national geoportal of the IGN. Scale: 1 km grid.

8. Refers to clandestine gold panners in the Brazilian Amazon and on the Guiana Shield.

9. Interview with the J2 Intelligence Bureau of the French Guiana Armed Forces High Command (COMSUP), Cayenne, May 18, 2018.

National Commission for Place Names admits the error, but considers that denominations coming from a sovereign ministry are generally considered as fully legitimate[10], which is probably why they were not questioned – incorrectly.

### 10.3.3. *Soliciting and then circumventing the state to make indigenous place names visible*

Contrary to the popular belief widely promoted by the aforementioned IGN maps, the Guianese forest is not a virgin territory with a few rivers named by rare explorers of past centuries and a few placers identified by wildcat gold miners on the borders. As evidenced by the abundant archaeological remains throughout the Amazon and the Guiana Shield, the Amazon rainforest began to be explored at least 13,000 years ago and is therefore far from pristine (Rostain 2013). The interior is still today the place of life for indigenous peoples and Black Maroon peoples, settled in villages along the rivers and around areas where shifting cultivation is practiced using slash-and-burn techniques, as well as fishing and hunting. The way of life of communities in southern Guiana is based on a *resource territory* that caters to the needs of these forest communities in many ways. Consequently, each cove, each plot of land specific to the culture, each inselberg has its own character, a history and a place name which carries part of this memory and of these contemporary practices. In Guiana, the memory of indigenous communities is inscribed in the histories of the territory, and even more so in place names. Place names organize space and socialize it.

It is to preserve this living cultural heritage and to facilitate the recognition of the appropriation of this territory by indigenous communities that the Guiana Amazonian Park (PAG) launched a vast participatory mapping project which was completed in 2014, for the Teko and Wayãpi communities on the southern border of French Guiana. By capitalizing on various surveys carried out over half a century and by performing, from 2009 to 2011, systematic GPS surveys, the PAG in partnership with the CNRS was able to draw up maps of the municipality of Camopi comprising nearly 600 different place names transcribed in the Wayãpi and Teko languages. The Wayãpi and the Teko are two indigenous peoples from the large Tupi-Guarani family, widespread in Brazil. Their long and intimate knowledge of waterfalls, passes, hunting paths, gathering and fishing areas, former villages… is expressed in the very large number of place names whose territory is enmeshed (Grenand et al. 2017).

10. Interview with Pierre Jaillard, President of the National Commission for Place Names, Paris, February 21, 2018.

To have this approach and this indigenous knowledge recognized, the PAG, at the request of the indigenous communities of Camopi, structured and standardized a toponymic database which it then undertook to distribute via official channels. It thus integrated this database into the GeoGuyane site[11], a public platform administered by the services of the French State. As such, it represents the Guianese side of the European spatial data infrastructure implemented since 2007 in the framework of the INSPIRE directive. The latter aims to promote the reuse of institutional data by encouraging their dissemination and promoting access to them by defining technical specifications. In France, the national Geocatalogue[12] provides the intermediary between the European portal and the regional infrastructures which are themselves fed by local infrastructures. This interlocking of platforms, making it possible to trace data from the local level to the European level, must operate based on *harvesting* techniques, which *harvest* the data as soon as they are disseminated on standardized infrastructures. Integrating its toponymic database into GéoGuyane was therefore for the PAG the symbol of official recognition of indigenous place names, by ensuring its distribution on national and European portals and, at the same time, multiplying the possibilities of reusing this data. But in practice, the operation turned out to be much more perilous…

#### 10.3.3.1. *Refusal to integrate indigenous place names: a question of sovereignty*

Six months after the transmission of the data, the Amazonian Park's request received a negative opinion from the official authorities. This refusal of integration was motivated by a double reason. In its letters addressed to the PAG, the IGN explains that 350 place names are in "legitimacy conflicts": these are places already named by another source and which require two different sources to be integrated. The Amazonian Park points to the dozens of people, scientists and inhabitants who have contributed to this participatory mapping. But for IGN, it is the same instance, the same group of inhabitants and therefore the same *source*. We see here the weight of the past and the difficulty of questioning the names given by the explorer colonists, or by certain soldiers unwittingly. The National Commission for Place Names also considers that uniqueness of sources does not pose a problem, in the case of the database transmitted by the military, because it comes from a sovereign Ministry, as opposed to the database from the Guiana Amazonian Park. In addition, 324 place names are in "technical conflicts": characters specific to indigenous languages are indicated here which do not respect international standards. This last argument refers to a phoneme common to Teko and Wayãpi which does not find a graphic translation in French. It is a vowel between "i" and "u" denoted by

11. https://www.geoguyane.fr/.

12. http://www.geocatalogue.fr/.

convention by a “crossed i with tilde” or “ɨ̃”. The tilde is a diacritical mark; it accompanies a letter to modify its pronunciation, in this case to nasalize the vowel. IGN considers that this notation does not comply with the standards in force. The PAG contacted the UNICODE international standardization consortium, which sets the computer standard allowing exchanges of texts in different languages, and demonstrated with the help of this consortium that the “crossed i with tilde” could be used in all computer systems that comply with the international standard.

This technical argument over, discussions with those responsible for place names in Paris then revealed other considerations, less technical and more political. Seen from Paris, it is difficult to conceive that there could be so many indigenous place names; it is difficult to imagine a greater toponymic density in Camopi than in Cayenne and, above all, the place names of the border municipalities raise a strong issue of sovereignty: “The fact of naming a place may be used one day to establish a territorial claim. There is therefore a real issue of sovereignty around place names, in particular in these border areas”[13].

#### 10.3.3.2. *Bypassing the State: the geoweb as a weapon of resistance*

Faced with this institutional blockage, the geoweb, through the many online mapping platforms that develop there, could be considered as a bypass for the dissemination of toponymic registers with high heritage value but not officially recognized. OpenStreetMap quickly emerged as a credible alternative. From a technical point of view, the integration of the “crossed i with tilde” is not a problem, and the OSM data model makes it possible to manage multilingualism from the outset.

In addition to these technical considerations, the growing institutionalization of OSM now makes it possible to make it a tool for disseminating and recognizing work that is already accomplished. The ease of access to data, the flows of which are freely open and interoperable, considerably broadens the potential for dissemination. Whereas, until now, maps were considered authoritative because of their official character, they now acquire a new form of authority thanks to the ease of access and reuse of their spatial data. From this point of view, OSM seems to be a serious competitor for IGN, even among the French ministries. Indeed, although not accepted into the national database, indigenous place names have gradually been incorporated into maps produced by private organizations, NGOs and local government authorities. Among the most striking examples, we may cite hydrological maps produced by a department of the Prefecture, and health care

13. Interview with Pierre Jaillard, President of the National Commission for Place Names, Paris, February 21, 2018.

access maps distributed by the Regional Health Agency (ARS). Faced with multiple delays and blockages on the official path, the Teko and Wayãpi communities of French Guiana are charting a parallel path which they control, and which is gradually spreading, betting on the fact that it will eventually prevail (Noucher 2020b).

## 10.4. Research agenda: when the geoweb brings place names into the era of post-sovereignty

### 10.4.1. *Questioning the paradoxical promises of the geoweb*

The analysis of the control of place names in French Guiana, through the deconstruction of the cartographic myths of the past and the IGN maps of today, makes it possible to question the way in which the French State seems to extend the colonial legacy and to highlight the power games around the control of territorial denominations. Putting it into historical perspective also makes it possible to insist on the important weight of imaginaries in the (non-)renewal of Guianese place names. This study finally reveals that, if the inscription of local place names on the maps published by the allochthonous power may become a potentially effective method of reclaiming indigenous identity, in French Guiana, this inscription comes up against issues of informational sovereignty. However, in the face of socio-technological changes, and the emergence of players with global dimensions – from multinationals like Google to communities supporting digital commons like OpenStreetMap – cartographic deregulation is accelerating. Counter-toponyms are spreading on the Web, and the authority of institutional data is gradually being challenged. The geoweb then bears a paradoxical promise: it offers access to a diversity of representations of space while highlighting, through various devices for controlling and prioritizing information flows, the most formally uniform content.

Thus, in the age of the Internet, it now seems necessary to deconstruct the process of regulation and circulation of the various toponymic databases, beyond official maps alone. The way in which these databases are disseminated, appropriated and reappropriated by different spheres of actors conditions their effective use and ultimately their authority. Confronted with the geoweb, the institutions, mainly national ones, which had until then designed, produced and supplied the cartographic and toponymic figures of reference, are challenged and competed with by actors with very diverse statuses and motivations, basing their legitimacy on the proximity allowed by the dissemination of digital technologies. The visibility or invisibility of different choices of place name requires a deciphering of the games of authority which subsequently take place (also) on the Internet. Without replacing *place naming studies* but on the contrary, by seeking to extend them, I invite, through this contribution, research on political toponymy to multiply empirical studies and critical analyses of the processes of (de-)regulation of digital data flows. By positioning itself downstream from the creation of place names, this involves an engagement with real *place name flow studies*.

### 10.4.2. *Deconstructing data flows*

Inscribed in the ideology of "informational liberalism" (Loveluck 2015), data flows have become an essential figure in contemporary modes of information management. They are accompanied by notions of decentralization, network or even fluidity which come to oppose the centralizing rigidity of social and political entities, first and foremost the state. Informational liberalism is a reformulation of liberal political theory based on the widespread circulation of digital content within now-interconnected computer networks. Information must be free and systems standardized so that these flows of information solve multiple problems and allow innovative services to emerge. The whole concept is therefore based on a common utopia, a "technological sublime" (Nye 1994) where the circulation of information is promoted as a vector of democracy and prosperity. These techno-scientific promises deserve to be questioned, as highlighted by our case study on French Guiana.

Technically, it is at the same time the efforts to standardize information systems, the lowering of communication costs and the improvement in the performance of communication infrastructures that have enabled the development of new computerized tools aimed at facilitating the aggregation, the fusion, the intersection, the hybridization of sources of diverse origins. While the production of toponymic databases has long functioned "in silos" with the multiplication, in the 1990s, of sectoral geographic information systems (one GIS per territory and per theme), the interoperability which accompanies the development of the Web today accelerates the networking of information. The simultaneous assembly of many data infrastructures and their interconnection at a planetary scale allow the rapid constitution of a growing mass of information which, although dispersed over different servers, can easily be gathered by a user and processed simultaneously. Information system experts talk about *orchestration* to discuss the control operations of these data flows. These are sometimes thought of as obvious, as automation makes them invisible. However, it is precisely these logics of orchestration that critical toponymy should grasp today. For spatial data to be recognized by these systems, for them to circulate and be able to ultimately appear on a map, they must meet increasingly stringent standards. However, nothing is neutral in the choices of standardization of data flows: they are guided by principles, objectives, rules, in short, intentionalities that are to be deciphered. The management of the circulation of spatial data appears as a new form of control which operates a homogenization and a filtering of representations, gradually becoming a tool for the dissemination of unequivocal representations of the territory, where alternative representations to those of the State or the giants of the Web struggle to make their way. Henri Desbois underlines this paradox: "It is surprising to note that in our time when GIScience and the geoweb have led to an inflation of cartographic production probably comparable in its magnitude to what Europe knew during the distribution of the first printed atlases (Woodward 2007), we often all look at the same maps" (Desbois 2018, p. 321). Thus, collective visions are constructed and formatted which deserve to be discussed. Specific geographical methods must then be invented to contribute to the understanding of these decentralization processes, which in no way erase the effects of domination creating or reinforcing hegemonic positions, of which Google is undoubtedly the most emblematic example.

### 10.4.3. *Opening algorithmic black boxes*

The challenge is therefore daunting because exploring the socio-technical thickness of the infrastructures controlling these data flows is to confront real algorithmic black boxes. Concerns around the opacity of algorithms are widely shared in the scientific community (O'Neil 2016). They lead to the development of algorithmic studies with very varied approaches, as evidenced by the dynamism of the journal *Big Data & Society* in which multidisciplinary content (computing, information and communication sciences, law, sociology, geography, etc.) abounds. Each of the disciplines that mobilize around the subject tends to extend and renew its theoretical and methodological anchor points in contact with the digital devices studied. In geography, critical cartography has been deciphering, for several decades, from a cultural and political point of view, the construction of maps and their socio-spatial effects. My proposal is resolutely in line with the foundations of this current and its extensions around critical GIS (Schuurman 2000). But it also seeks to go beyond this anchoring: taking an interest in algorithmic black boxes in a context of post-cartographic sovereignty requires, in fact, considering new strategies. As Dodge and Perkins (2015) point out, it is undoubtedly much more complex to deconstruct current cartographic devices where spheres of various actors intertwine, where multiple technologies are aggregated and where data flows intersect from various sources, than the devices of yesteryear where cartographic power was held by a few rare experts. While Harley's studies center on deciphering the authority of maps and their overarching and dominating nature, today questions of data circulation are central. To gradually supplement the textual reading of the product by focusing more on the data networking process, new approaches may be imagined and developed. Stimulating research perspectives then open up to avoid abstract propositions and engage in embodied approaches. It seems essential to combine critical cartography with, on the one hand, the Geographic Information Science (GIScience), to be in contact with the designers of the systems studied, and with, on the other hand, Science & Technology Studies (STS), to understand both the way in which practices are constrained and the way in which uses adapt. Such a disciplinary alliance would make it possible, in the words of Rob Kitchin, "to unpack the entire socio-technical assembly of algorithms"[14] (Kitchin 2017, p. 25), without overestimating the relevance of technology alone.

Despite their gradual removal from the cartographic *stage*, flows of geographic data act, *backstage*, as vectors for filtering representations judged suitable or not for circulation. Also, in the same way as *place names studies* question the processes of creation of place names from a political perspective, this contribution was intended as a plea for the development of empirical interdisciplinary studies that make visible the issues of the circulation of place names, which currently contribute in a meaningful way to their imposition or omission.

## 10.5. References

Desbois, H. (2015a). *Les mesures du territoires : aspects techniques, politiques et culturels des mutations de la carte topographique*. Presses universitaires de l'ENSSIB, Villeurbanne.

Desbois, H. (2015b). La carte et le territoire à l'ère numérique. *Socio*, 4, 39–60.

Dodge, M. and Perkins, C. (2015). Reflecting on J.B. Harley's influence and what he missed in "deconstructing the map". *Cartographica*, 50(1), 37–40.

Douzet, F. (2020). Du cyberespace à la datasphère. Enjeux stratégiques de la révolution numérique. *Hérodote*, 177/178, 3–15.

Douzet, F. and Desforges, A. (1998). Du cyberespace à la datasphère. Le nouveau front pionnier de la géographie, *Netcom*, 32(1/2), 87–108.

Faure-Muntain, V. (2018). Les données géographiques souveraines. Rapport au Gouvernement, July.

Gautreau P. (2021). *La Pachamama en bases de données. Géographie politique de l'information environnementale contemporaine*. Éditions de l'IHEAL, Aubervilliers.

Granger, S. (2011). Contesté franco-brésilien : enjeux et conséquences d'un conflit oublié entre la France et le Brésil. *Outremers*, 98(372–373), 157–177.

14. Author's translation.

Grenand, P., Grenand, F., Joubert P., Davy, D. (2017). Pour une histoire de la cartographie des territoires teko et wayãpi (Commune de Camopi, Guyane française). *Revue d'ethnoécologie*, 11 [Online]. Available at: https://journals.openedition.org/ethnoecologie/3007.

Grumbach, S. (2020). Gouvernance numérique et changement climatique. *Hérodote*, 177/178, 17–31.

Harley, J.B. (1989). Deconstructing the map. *Cartographica: The International Journal for Geographic Information and Geovisualization*, 26(2), 1–20.

Jacob, C. (2005). *The Sovereign Map. Theoretical Approaches in Cartography Throughout History*. University of Chicago Press.

Joliveau, T., Noucher, M., Roche, S. (2013). La cartographie 2.0, vers une approche critique d'un nouveau régime cartographique. *L'Information Géographique*, 77(4), 29–46.

Kitchin, R (2017). Thinking critically about and researching algorithms. *Information, Communication and Society*, 20(1), 1–14.

Kwan, M.-P. (2002). Is GIS for women? Reflections on the critical discourse in the 1990s. *Gender, Place and Culture: A Journal of Feminist Geography*, 9(3), 271–279.

Laboulais-Lesage, I. (ed.) (2004). *Combler les blancs de la carte. Modalités et enjeux de la construction des savoirs géographiques (XVII^e^–XX^e^ siècles)*. Presses universitaires de Strasbourg.

Lézy, E. (2000). *Guyane, Guyanes, une géographie "sauvage" de l'Orénoque à l'Amazone*. Belin, Paris.

Lieselot, L., Ooms, K., De Maeyer, P. (2020). The influence of map projections on people's global-scale cognitive map: A worldwide study. *ISPRS International Journal of Geo-Information*, 9(4), 196.

Loil, P. (2011). E793 : la nouvelle cartographie de la Guyane. *Echanger, mensuel interne de l'IGN*, 228.

Loveluck, B. (2015). *Réseaux, libertés et contrôle. Une généalogie politique d'internet*. Armand Colin, Paris.

Mercier, G. (2009). La géographie de Paul Vidal de La Blache face au litige guyanais : la science à l'épreuve de la justice. *Annales de Géographie*, 667, 294–317.

Morrachini, V. and Comptour, M. (2020). Toponymie des quartiers de Saint-Laurent du Maroni. In *Atlas critique de la Guyane*, Noucher, M. and Polidori, L. (eds). CNRS Éditions, Paris.

Noucher, M. (2017). *Les Petites Cartes du Web. Approches critique des nouvelles fabriques cartographiques*. Édition Rue d'Ulm, Presses de l'École normale supérieure, Paris.

Noucher, M. (2019). Information géographique volontaire, notice de l'encyclopédie en ligne. *Hypergeo* [Online]. Available at: https://hypergeo.eu/information-geographique-volontaire/.

Noucher, M. (2020a). La carte au 50 000e : de la couverture homogène au patchwork de données. In *Atlas Critique de la Guyane*, Noucher, M. and Polidori, L. (eds). Éditions CNRS, Paris.

Noucher, M. (2020b). The place names of French Guiana in the face of the geoweb: Between data sovereignty, indigenous knowledge, and cartographic deregulation. *Cartographica: The International Journal for Geographic Information and Geovisualisation*, 55(1), 15–28.

Nye, D.E. (1994). *American Technological Sublime*. MIT Press, Cambridge.

O'Neil, C. (2016). *Weapons of Math Destruction: How Big Data Increases Inequality and Threatens Democracy*. Crown, New York.

Perdana, A.P. and Ostermann, F.O. (2019). Eliciting knowledge on technical and legal aspects of participatory toponym handling. *ISPRS International Journal of Geo-Information*, 8(11), 500.

Quinn, S. (2016). A geolinguistic approach for comprehending local influence in OpenStreetMap. *Cartographica: The International Journal for Geographic Information and Geovisualization*, 51(2), 67–83.

Rambaldi, G. (2005). Participatory 3-D modelling. *Encyclopedia of Developing Regional Communities with Information Communication Technology*, Idea Group Reference, 538–543.

Regourd, F. (2004). Coloniser les blancs de la carte. Quelques réflexions sur le vide cartographique dans le contexte colonial français de l'Ancien Régime (Guyanes, Antilles françaises, XVIIe–XVIIIe siècles). *Combler les blancs de la carte. Modalités et enjeux de la construction des savoirs géographiques (XVIIe–XXe siècles)*, Laboulais-Lesage, I. (ed.). Presses Universitaires de Strasbourg, 221–242.

Rostain, S. (2013). *Islands in the Forest: Landscape Management in Pre-Columbian Amazonia*. Routledge, New York.

Rouard de Card, E. (1897). *Le différend franco-brésilien relatif à la délimitation des Guyanes*. A. Pedone Éditeur, Paris.

de Vilhena Silva G. (2017). Litiges transfrontaliers sur le plateau des Guyanes, enjeux géopolitiques à l'interface des mondes amazoniens et caribéens, *L'Espace politique*, 31(1), April 18 [Online]. Accessed 20 August 2022 at: http://journals.openedition.org/espacepolitique/4242.

Walker, R., Jojola T., Natcher, D. (eds) (2013). *Reclaiming Indigenous Planning*. McGill-Queen's University Press, Montreal.

Woodward, D. (ed.) (2007). *The History of Cartography, vol. 3: Cartography in the European Renaissance*. University of Chicago Press.

# 11

# What Africa Might Contribute to Critical Toponymy

**Michel BEN ARROUS**[1] **and Liora BIGON**[2]
[1] *Gaston Berger University, Saint-Louis, Senegal*
[2] *Holon Institute of Technology, Israel*

> The fathers had left Darou Khoudoss in Baol to found Darou Khoudoss in Saloum. The sons extended the route of the fathers to Fuladu and created another Darou Khoudoss in the forest of Pata. Like them, they responded to the call of the marabouts and cleared land to grow groundnuts. It is not one or two villages that they created, illegally, in the classified forest, but twenty or more. [...] Our *kaabunke* grandparents also, when they left Portuguese Guinea for this corner of Casamance, followed marabouts. They established their own villages on the edge of the forests, and then rings of other villages around the first. They had, like the Saloum-Saloum today, powerful political and brotherhood support. The only ones who don't are the Fulani *jiyaabe* [former slaves]. They were the ones who won the independence of Fuladu. They left the forest of Pata after naming each pond, each track, and as a last resort each *sinthiane*. (Emballo 2007, p. 8)

The intention of this chapter is to take critical toponymy out of its comfort zone. By comfort zone, we mean a research habitus that covers a set of approaches, priority objects and interpretative schemes which make critical toponymy studies "a bit too predictable" (Rose-Redwood et al. 2010, p. 466), repetitive and, above all, excessively reductive in African situations. Critical toponymy (hereafter CT) was built around a broad line of questioning: What is it that the naming of places *can say*

*The Politics of Place Naming*,
coordinated by Frédéric GIRAUT and Myriam HOUSSAY-HOLZSCHUCH. © ISTE Ltd 2022.

about the space of societies and the power relations unfolding there? In practice, however, the field has narrowed around official place naming alone, with a very strong focus on the ideological strategies of rulers and marginal treatment of the toponymic practices of the governed.

We offer in the following pages a critique of this narrowing. From there, we sketch the outlines of a possible renewal of CT based on certain major characteristics of place naming in contemporary Africa. The most remarkable of these, and also the most neglected, is the diversity of toponymic forms – in other words, the lack of hegemony of official forms.

The epigraph of this chapter illustrates a situation of multiple place naming in Senegal. If one focuses only on the official place names (Darou Khoudoss, Pata), one will only grasp the advance of a frontier of agricultural colonization, generating tensions between migrants and so-called indigenous populations. Other toponymic forms provide information about deeper dynamics. These are firstly the old forms, with no legal existence in contemporary Senegal. Fuladu, Baol, Saloum, ancient kingdoms now defeated, remain the usual frameworks of collective identification, and therefore of mobilization around polarizing issues. They testify to different relationships with space, based for some on the groundnut economy and the spread of a brotherhood-based Islam, for others on defense of a territory conquered by arms[1]. The microtoponymy of the lived space, also "informal", indicates the most disputed places – water points, cattle tracks and *sinthiane*, which designate here the limits of maximum expansion of terroirs. Finally, Casamance remains another area of reference, although a territorial reform in 1984 formally deleted its name from the list of regions of Senegal.

Multi-toponymy suggests other logics than official toponymy alone. It reflects political imaginations that draw from different levels of historical depth, the activation of a memory of conflicts and slavery, games of alliances between the state and networks of brotherhoods. The state's inability to impose its own legality in a classified forest is therefore no longer necessarily read as a sign of weakness. It can also be seen as the actualization of a project of political engineering inherited from the colonial period, aimed at normalizing, through the creation of maraboutic

1. Fuladu ("land of the Fulani" in Mandinka) was at the end of the 19th century a kingdom founded by former slaves, seceding from the enslaving kingdom of Kaabu. The latter was a periphery of the former Mali Empire, then, when Mali broke up in the middle of the 16th century, an independent kingdom. Saloum and Baol were kingdoms issuing from the empire of Djolof, itself a vassal of the empire of Mali. The point on which we insist is the permanence of these unofficial toponyms in current usage and their capacity for signifying collective identities: to be Saloum-Saloum (or Baol-Baol) is to be "native" of Saloum (or from Baol). Emballo defines himself as *kaabunke*, that is, "native" of Kaabu.

villages, a turbulent region. Our intention here is not to present in detail a particular case that is well documented elsewhere (Sidibé 2002; N'Gaïdé 2009; N'Gaïdé 2015). Rather, it is to underline the diversity of toponymic forms – and to observe that analyses may vary greatly depending on whether one focuses on a single form or examines their articulations.

The simplest multi-toponymic situation, the least foreign to the usual CT reasoning, is that where the same place receives different more or less competing names. But multi-toponymy is above all about heterogeneous naming systems, which are attached to different categories of places and give them different meanings. The question then is identifying the most heuristic places or namescapes, those that *can say* the most about the space of societies. This presupposes taking an interest, much more seriously than CT has done so far, in the reception of names, in their differentiated uses depending on context. In all these respects, what Africa might contribute to CT is a programmatic, methodological and theoretical renewal.

Generalizations obviously remain difficult at the level of such a vast continent. So, we must say a word about the place from which we are writing: Dakar. In a certain way, it is a coincidence. Dakar is simply the city we know best – one of us having spent most of his life there, the other having devoted her thesis to it. We are aware that we would probably write differently from Harare or Abidjan. But one always writes and theorizes from somewhere, and might as well take advantage of it. The advantage, in this case, is the way in which Dakar evades common approaches. The Senegalese capital, former capital of French West Africa (AOF), is indeed not the city best suited to observe a spectacular reversal of colonial toponymy. From this point of view, it serves rather as a counter-example, an archetype of the city where nothing interesting happens for CT. The continual renewal of its multi-toponymy, however, testifies to a proliferation of counter-hegemonic practices. It is the vigor of these practices, and the knowledge/power relationships they inscribe in the urban space, that guide our thinking. We also mobilize the experience of other places and sites of study, as well as a corpus of African studies that is sensitive to the articulation of different toponymic forms. On the contrary, we pass more quickly over conventional studies, focused on official place naming alone, which to date constitute the bulk of the literature.

## 11.1. Official toponymy and others

In 1993, there appeared in the United States *Africa and the Disciplines*. Co-edited by political scientists Robert H. Bates and Jean O'Barr and philosopher Valentin-Yves Mudimbe, the book examined the contribution of African studies to the production of knowledge in the social sciences and the humanities. The book was intended as a defense of Africanism in the American university system. It took place in a context marked by the marginalization, since the end of the Cold War, of area studies in general (Cumings 1998). Research fields were being reconfigured, access to funding was becoming more and more competitive, and the sustainability of African studies departments, in particular, seemed threatened (Guyer 1995). In this anxiety-provoking context, *Africa and the Disciplines* underlined the contributions of Africanism to disciplines considered to be fundamental (core disciplines) – anthropology, economics, political science, philosophy, history and art history, comparative literature. The reasoning was clever: if African studies were useful to these disciplines, then their presence on American campuses would be worth maintaining.

*Africa and the Disciplines* was well received in the United States and Europe, where it is still required reading in several African studies departments. African criticism, on the other hand, remained "hardly audible" (Mamdani 1997, p. 149). It focused on fundamental points directly concerning the African developments of CT: the nature of the contributions of research in Africa, the validity of the theoretical models to which they contribute, and the very notion of research *in* Africa – as opposed to research *from* Africa (Hountondji 1994). A review published in a journal of CODESRIA, one of Africa's leading social science research centers, deplored the "tendency" of Robert Bates and his co-authors to "assume that Africa's contribution to the disciplines lies merely in its peculiarities", real or supposed, and "capacity to provide validation to theories already developed elsewhere" (Zeleza 1994, p. 182). The Africa of *Africa and the Disciplines* was a repository of empirical materials, rather than an "epistemic site" in its own right, from which it would be legitimate to interpret the reality of both Africa and the world (Ndlovu-Gatsheni 2018).

This perspective remains. It continues to inform the most diverse fields of knowledge, including toponymy. African works of CT have grown rapidly over the past decade[2]. But if they were to be the subject of a chapter in an expanded edition of *Africa and the Disciplines*, what would we find there? Mainly "validations" of models "developed elsewhere", and little questioning. We would find the same focus as in Europe or America on the commemorative dimensions and the political

2. Two bibliographies published 12 years apart testify to this rapid development. The first (Batoma 2006) is almost exclusively composed of onomastic approaches. Critical approaches are largely predominant in the second (Neotopo 2018), with post-apartheid South Africa occupying a prominent place. Since the coming to power of the *African National Congress* in 1994, the country has embarked on an unprecedented program of toponymic overhaul, at all scales (streets and neighborhoods, towns and villages, provinces and districts, natural sites and monuments) and in the 11 languages recognized by the new constitution. Monographs on this institutional process are numerous. For provisional summaries, see Ndletyana (2012), Mavungu (2016), and Adebanwi (2017).

instrumentations of official place naming – whereas here it is only one toponymic form among others, with references drawing on other registers. We would also observe somewhat strange situations: places without an official name, places named without being precisely located, places named differently depending on the context, toponymic manipulations that provoke no passion, cities without an addressing system and addressing systems that no one uses. But how to account for it when the idea prevails that each place must correspond to a unique name, and when we postulate the hegemony of official names, their knowledge and their appropriation in usage, or when we credit their political instrumentation with effectiveness that it does not always possess?

At least part of the usual toponymic reasoning is based on assumptions that do not hold true in African situations. We identify some of them in the following pages. Their withdrawal would lead to reshaping the conceptual apparatus of CT, in its most fragile intuitions and generalizations, but might also extend its research horizons and its explanatory scope. This would be an initial significant contribution. Failing this, the "peculiarities of Africa" will continue to be dealt with (or ignored) relative to an external order, constructed as a standard of normality.

CT themes have diversified over the past three decades, as have its sites of study, at the same time as it has asserted itself as a legitimate field of research. The unification of the field took place around the epithet "critical", the common denominator of a set of perspectives on the naming of places – toponymy as a technology of power, as writing or rewriting of history, as arena of symbolic struggle, as an issue of social justice, in particular (Vuolteenaho and Berg 2009; Rose-Redwood et al. 2010). However, CT remains an emerging – in other words, malleable – field, whose lines of research, issues and theorizations are constructed and renewed, iteratively, through empirical studies, conceptual intuitions and their testing. It remains to make the most of each site of study.

In this respect, the relevant objects are not the same everywhere. Official names, in particular street names and their changes, were established as privileged objects very early on. This was consistent with the places of emergence of CT, on mainly European and North American bases. The renaming of streets was a leading indicator of the political and ideological transformations taking place in the former Eastern Bloc. Elsewhere in Europe or North America, the imposition, reclaiming or erasing of specific place names also provided information about the transformation of power relations between states and their ethnolinguistic or regional minorities. But official toponymy is not always so instrumental, nor even so present. According to estimates by UN-Habitat (2016), the proportion of urban residents living in so-called informal settlements – where streets have no legal names – is around one in five inhabitants in

Latin America, one in three in Asia, and more than one in two in Africa. That is a lot of people to ignore if we stick to the study of only official place names.

Outside of "informal" neighborhoods, named or numbered streets are absent from entire cities, or so rare that even the administration and the service industry locate residents or customers by landmarks (Figure 11.1). Pikine, the second most populous city in Senegal, has only four named streets (Figure 11.2). Wayfinding by landmarks prevails even when a conventional addressing system exists. In practice, it is not the streets that are named but significant places, sections of streets or intersections – like the Carrefour J'ai raté ma vie ("I wasted my life" crossroads), the Carrefour des Trois Voleurs (Three Thieves crossroads), the Carrefour des Trois Boutiques (Three Shops crossroads) or Bépanda Double Balle (Bépanda Double Gunshot) to stick to place names well known to the inhabitants of Douala in Cameroon (Amougou Mbarga 2013).

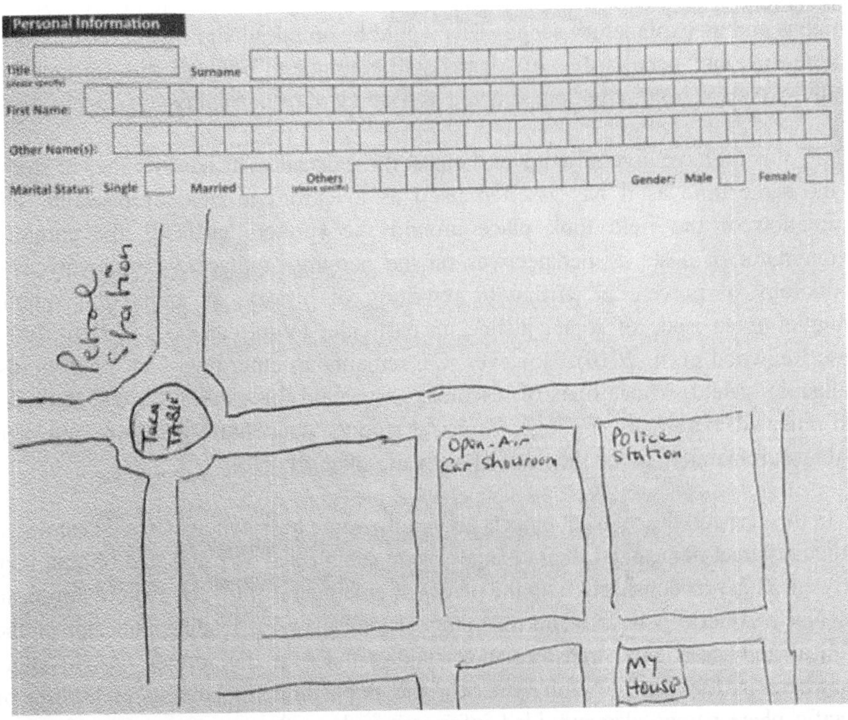


**Figure 11.1.** *Form for opening an account in a banking establishment in Banjul, The Gambia. A white space is provided to allow customers without an official address to draw a sketch of the location of their home. This form has recently been revised, replacing the white space with a "nearest bus stop or landmark" box (source: Ade Daramy 2019)*

**Figure 11.2.** *Aerial view of Pikine on the outskirts of Dakar, Senegal. A satellite town created in the 1950s to rehouse a population that had been* déguerpie *(i.e. displaced by the authorities from the working-class neighborhoods of Dakar), Pikine is now home to more than a million inhabitants. Since its creation, it has had four named streets. The rest have been numbered recently, without modifying the usual practices of wayfinding based on retail shops, their owners or the residence of local personalities (photo: Liora Bigon, January 2018)*

Multiple naming systems coexist, designating the same places differently or giving names to others which the official toponymy ignores. CTs almost exclusive focus on the latter did not arise from a reflexive process. It is not the result of an internal debate in the field, or from a well-argued theoretical position. Rather, it reflects a dominant propensity, a largely unconscious inclination, followed by a field of research that was built without particular attention to African namescapes. By replacing multi-toponymy and its uses at the heart of the analysis, we perceive the interest, and also the possibility of a heuristic reversal: contributing *from* Africa to a constructive criticism of CT, rather than validating, *in* Africa, pre-established grids of analysis. If the latter ignore the experience of African societies with multiple forms of naming, it is perhaps not, in fact, the African experience that is outside the norm; it is rather the analysis grids themselves which, upon examination, prove to be unduly prescriptive, or insufficiently universal.

In CT as in other fields of knowledge, the reduction of African situations to fit them into in pre-existing grids promotes stereotypes. It suffices, when societies diverge from models that we imagine to be universal, to invoke their “specificities” – which elsewhere would be more simply called contextual elements – or, what comes to the same thing, to disqualify them as irrelevant. Comparative approaches are then distorted from the outset, trapped between two symmetrical impasses: one that relegates Africa to an absolute singularity to explain that what is valid elsewhere is not verified there; the other which, in contrast, neglects the elements of non-conformity with the models that one seeks to validate. Between these two impasses, historicism draws a third, a teleological vision according to which Africa and other “non-conforming” regions of the world will lose, over time (Duara 1995; Chaudhuri 2004), their singular characteristics.

But it is obviously not enough to add a speculative dimension to a normative bias to better grasp the situations. Two recurrent themes of African CT studies are particularly impoverished by teleological approaches:

– The development of efficient addressing, that is, one that complies with international standards, is seen as a necessity for development in general. Arguments for this are numerous – improving urban governance, extending municipal tax revenues, stimulating e-commerce, making life easier for residents in multiple situations where an address seems necessary: enrolling a child in school, opening a bank account, calling an ambulance, etc. The World Bank financed a program in this direction in the 1990s, covering 52 cities in 15 countries (Farvacque-Vitkovic et al. 2005). The relative failure of the program, due to a lack of uptake by the residents of the towns concerned, is well documented. It is attributed to internal causes (heavy-handed operations, technocratic coldness, etc.), which are certainly real, but which tend to conceal the failure of other similar attempts. The lack of uptake of addressing, where it exists, is thus still perceived as a technical setback, rather than a recurring phenomenon whose social and political dimensions might be worth questioning. By reducing the fabric of lived space to the level of an annoyance, by keeping it outside the frameworks of analysis, we refrain from thinking about policies of addressing from a critical perspective. Their lack of social legitimacy nevertheless constitutes a powerful indicator of the tensions by which the lived space and the planned city, desired but never arriving, mutually inform and transform each other – differently depending on the context, but always by testing the authority of rulers (Bertrand 2001; Wade 2009).

– Another drastic reduction of the field of observation is the "decolonization" of street names. This is today one of the most extensive threads of CT studies in Africa. It consists mainly of monographs. The replacement of street names of colonial origin by names celebrating national heroes or great African figures is considered as an inevitable and necessary alignment of the symbolic order with the political order. The instrumentalization of street names by colonial then postcolonial rulers thus remains seen, in its continuity, as an unavoidable modality of political action. By necessity, these monographs focus on city centers – where there were, and still are, colonial street names to be replaced. On this scale, renamings contribute to the "decolonization of the urban landscape", its "nationalization" or its "Africanization" (Wanjiru 2018, pp. 66–91). However, town centers remain very particular fragments of landscape. They can only truly be interpreted at the level of urban landscapes as a whole. Only an overall picture allows us to grasp the uneven distribution of named streets – and the ensuing paradox. As laudable as the (re)naming efforts are, their concentration in city centers (or other politically valuable neighborhoods) contributes in practice to reproducing, updating or even accentuating spatial inequalities that result, essentially, from colonial urban planning. The focus on city centers also encourages us to consider the decolonization of street names as the alpha and omega of memory marking in postcolonial situations. It means ignoring the dynamics of societies, the continual renewal of their polarizations and collective mobilizations, and the diversity of memories and counter-memories that are involved in the urban space or seek to be involved in it (Ndletyana 2012; Wanjiru 2018, pp. 94–128; Bigon and Ben Arrous 2019). By keeping informal toponymies off our radar, we neglect that they constitute, on these questions, an important living archive.

CT is not immune to the question of comparativism. The diversification of its sites of study makes it necessary to formulate overall frameworks, which might bring together a growing body of work and authorize shared syntheses. It is still necessary to acquire relevant comparables, allowing heuristic comparisons without overly impoverishing the situations. The proposal that comes closest remains to this day a comparative framework based on the Foucauldian notion of *dispositif* (Giraut and Houssay-Holzschuch 2016). The framework is articulated around categories of actors, geopolitical contexts, and modes of naming places, among which we can identify preferential combinations. This framework proves convincing for deciphering the political logics of official (re)namings (Adebanwi 2017; Wanjiru 2018). However, it remains silent on the multiple toponymy that makes up the richness – and the interest – of African situations. At the cost of some adjustments, perhaps we might consider that here, several *dispositifs* operate simultaneously. But we would still fall short of what seems most necessary for the interpretation of African multi-toponymies: taking into account the *reception* of names (formal or informal), the practices and uses that mobilize them or turn away from them, and the knowledge/power relationships that are at stake in these positions. To our knowledge, there is no reason other than habit, that is, confinement to a routine, for sticking to the analysis of institutional logics when the reception and use of names may inform us, much more broadly, about the internal dynamics of societies.

On a methodological level, it is, of course, more convenient to stick to conventional names, those found in signs, maps or directories, than to study actually existing toponymic practices. The latter, however, remind us that if the naming of places is a universal human disposition, its forms vary, and so do its issues. Rather than evading this diversity, or rejecting it as a supposed African exception, it can be understood as an invitation to re-examine the conceptual bases of the analysis. The concepts of *place*, of *power* and of *power over space,* central in critical approaches, are not unequivocal. Questioning their meaning in specific contexts amounts to opening the "black box" of toponymic systems. What then becomes observable and comparable are no longer toponymic forms or even practices alone. It is also, in a much more fruitful way, the systems of values and references that these forms and these practices mobilize, the relationships to space upon which they are founded, and the social dynamics or the power relations that drive their transformations or their arrangements.

## 11.2. A problem of places

The naming of places encompasses two issues: *naming* and *places*. The first polarizes attention. It is clearly political, insofar as the choice of a name "brings together different actors, various projects, and often conflicting identity or functional representations, certain of which will ultimately be promoted" (Giraut et al. 2008, p. 98). The other term, *places*, is rarely problematized. The power to name begins, however, upstream, with the identification of what makes a place. What is being named? What is worth being named in the space of societies? The answers involve operations of classification, taxonomies that are more or less shared, accepted, contested, ignored. In these operations, too, power relations, possibilities of conflict, and possibilities of evasion are entangled.

What makes a place for one human group does not necessarily make a place for another. In its simplest definition, a place is a portion of space invested with collective meanings. A rock cliff, a water point, a remarkable tree – each makes a place differently for different social formations. In Tuareg country, for example, trees constitute the most important category of named places. A lone tree or a grove in the desert is not only a landmark for wayfinding, but also a milestone that punctuates the collective experience of travel and rest – a place in its own right (Bernus 1980)[3]. Other societies, from Yoruba country to the Great Lakes region, also conceive, construct, and designate singular trees or other landscape features as places. They are sometimes even "high places", inscribed in a founding narrative or sanctuary according to their importance for a cult, a belief, or more broadly for the moral economy of the societies in question (Luig and Von Oppen 1997; Anyinam 1999).

These and other places, as diverse as the historical experience of societies and their forms of existence in space, are integral to countless sensitive geographies, symbolic attachments and concrete territorializations. Official nomenclatures do not systematically list them. They do not identify every cave, every well, every rock inscribed in a cosmogony or in the living memory of a community. Their names only subsist in and are transmitted through usage, through the naming practices of those who recognize in them a part of their collective identities.

Figure 11.3 shows one of these high places without a formal toponym in Gao, in northern Mali. Like most cities in the Niger Bend – the great curve formed by the Niger River on the edge of the Sahara – Gao is a city of canals. The first, according to oral traditions, were dug in the 4th century by a fishing community led by Faran Maka Bôté. They were used to bring boats closer to the camps in times of flooding, allowed the development of agricultural plots and, it is said, housed the spirits of the water. Other canals would later bring water from the river to the palace of the Askia – the dynasty that reigned in the 16th century over the Songhai Empire. Most of these canals are now buried. Their exact location and their names remain known to everyone. Like many other place names absent from official nomenclatures, they continue to make sense "for those who live there" (Woodman 2015, p. 14).

If it does not name each significant place, the name-giving state shall, in contrast, establish categories of generic places, which may only have a place in name. Certain "places" vested with official toponyms remain "non-places" (Augé 1992), portions of space devoid of strong collective meanings[4]. The name, however

3. It is also a source of food, fodder and fuel, a reserve of pods and bark useful for making mats, ropes or ointments, a source of shade, and so on.

4. We use the neologism created by Marc Augé, but with a different meaning from his. Augé defines non-places through the *individual* experience of anonymity in spaces of circulation or transit. For our part, we insist on *collective* meanings or the lack thereof. Augé's non-places would be a product of what he calls "supermodernity", a notion that is superfluous to us. The non-places that interest us are rather a condition of possibility and a product of statist modernity.

tinged with ideology, is not enough to bring into existence as places all the administrative divisions, thoroughfares, sites and monuments over which the state exercises its power of naming. These become places when one or more human groups project a form of attachment onto them in connection with their collective experience. Otherwise, their naming rarely generates the great political passions of which CT is fond (Figure 11.4).

**Figure 11.3.** *The Tilafonso canal in Gao, Mali. This canal, now covered (and regularly uncovered), is said to have been dug 17 centuries ago by Faran Bôté, a central figure in the stories of the city's foundation. The surrounding neighborhood retains a toponymic trace of this history. Its name, Farandjireye, literally evokes "the era of Faran". The canal itself has formally become a numbered street, Rue 312, a toponym which no one uses. In practice, each inhabitant of Gao refers to the canals of the city by their proper names: Tilafonso, Gouroum, Aljana Bandia, etc. It is only under these names, unofficial but significant, that they constitute high places in the memory of the city (photo: Abdoulaye Ibrahim Touré, October 2017)*

**Figure 11.4.** *Call for suggestions for names for an urban densification project between Johannesburg and Soweto, South Africa. The South African constitution obliges the public authorities to consult the citizens on any question of common interest, including changes or creations of place names. The result is a large number of consultative processes, albeit with uneven participation. The (perceived) weakness of turnout serves as a recurring argument, especially for the white minority, to challenge the legitimacy of (re)namings in court (photo: Frédéric Giraut, May 2016)*

The name-giving state is historically a simplifying state, working in Africa as elsewhere to decipher the "social hieroglyph" and make it "legible" for the sake of administration (Scott 1998, p. 3). However, the differentiated trajectories of state modernity, in Europe and in its areas of imperial expansion, did not produce the same degree of simplification everywhere. We owe to sociologist Christian Topalov (2002) a very enlightening perspective, on a global scale, of urban divisions and the words that express them. His comparative analysis of urban "*anciens régimes*" reveals an extraordinary diversity of intertwined patterns, corresponding to multiple forms of power combining locally. Beyond combinations which are specific each time, we find from London to Tokyo and from Kairouan to Bombay the same profusion of urban divisions, registered in heterogeneous toponymic forms and corresponding to the concomitant exercise of distinct authorities – for example, a magistrate, a prelate, a distant lord or one of his nearby vassals, a guild of artisans or merchants, etc. African multi-toponymies can be understood in this respect, not as an ontological characteristic or an anomaly that CT analyses could safely neglect, but as an indicator of power structures co-present in the same place.

Trajectories of simplification, several centuries long in Europe, have been much shorter in Africa. Above all, they were brought here by the colonial state – a place-naming state, and also an ethnographer and cartographer state. The construction of ethnicity as a method of classifying the colonized, the codification of customs and languages as markers of identity, and the monitoring of the mobility of ethnic subjects in increasingly narrow territorial webs, oriented the colonial definition of real or supposed places (Worby 1994; Mamdani 1996; Costa Dias 2006). Finer analyses than can be permitted here would reveal important local or micro-local variations in how the colonial century affected the identification of places by colonized subjects themselves (Von Oppen 2003). But two commonly used toponymic categories, the street and the neighborhood, allow at least some general observations.

As far back as one goes in the history of African cities, named neighborhoods are found. Royal or clerical quarters, craftsmen's or foreigners' quarters, their names are known and stable enough to appear in local chronicles, oral history, or travelers' accounts. The same sources provide information on the layout of the streets or their appearance, but when they mention one in particular, it is in a descriptive way – "the street which passes in front of such a building", "the street which leads to such a place", etc.[5]. There are no named streets. The naming of streets is a colonial initiative in Africa. It occurs in conditions of total foreignness to the colonized societies. Neither its ideological repertoire, the showcase of the nationalisms of the European 19th century, nor its hierarchy of generic terms (boulevards, avenues, streets, alleys, etc.), which also matured in Europe, can "speak" to the colonized. On the contrary, the deployment of street names in well-defined neighborhoods refers to familiar forms of urban organization. The materials of the buildings, the architecture, the plantations of trees, confirm the singularity of a space which the colonial powers themselves distinguished by lexical creations – such as the "Plateau" in so-called French-speaking Africa (Sinou 2014). Ironically, street naming then only makes sense, and place, as an element of dividing the city into neighborhoods.

In his *Seeing Like a State*, Scott contrasts the ability of the simplifying state to "standardize" the social world with the ability of societies to "modify, subvert, block and even overturn the categories" imposed on them (1998, p. 49). But Scott himself simplifies a lot. His empathy for "the local", "the art of the locality" or "the resistance of the local" leads him to neglect the articulation between the simplifying interventions of the state and the self-transformation of the "local" or of the sense of place (Geschiere 1998; Ben Arrous 2009). The literature on the implication of

5. These descriptive forms remain in wide use today (Dorier-Apprill and Van den Avenne 2002; Cumbe 2016; Mokwena 2020).

colonial ethno-linguistic classifications in the formalization, territorialization and sometimes postcolonial hardening of community identities is immense (Amselle 1985; Bayart 1996; Mamdani 1996; Marshall-Fratani 2006; Kam Kah 2019). Toponymic analysis must open itself to questions of the reception and use of names in order to perceive these complex dynamics, particularly in urban areas.

Neighborhoods remain today the "priority urban frame of identity reference" (Wade 2009, p. 331). Their names mean something different depending on the state of inter-community relations. When these become tense, they may be perceived as ethnic markers, simply because of the language used[6]. They then communicate a form of autochthony, a prior history in the places, which constitutes a form of symbolic capital capable of supporting claims to specific rights. Issues of autochthony are found in operations of addressing, where the setting of neighborhood boundaries is generally more disputed than the names of streets. The articulation of ethnic codifications and urban divisions also contributes to modifying the rules of reception of migrants by establishing signposted routes; newcomers will be received according to their origin in such and such a district, named and known as a reception district reserved for a particular group (Onguéné Essono 2018)[7]. It should also be noted that the colonial institution of so-called traditional chiefdoms, often carried over into independence, has complicated, rather than simplified, the administration of cities (Mouiche 2013). The authority of traditional chiefs never really fits in with the interlocking grids of neighborhoods and their subdivisions. The result is a "heterogeneous administration" (Amougou Mbarga 2013, p. 204), and also additional possibilities of recourse for citizens in the event of a dispute.

Multi-toponymic landscapes have a fluctuating character. Societies or different groups within them sometimes favor one form of naming, sometimes another, depending on the particular circumstances and resources offered by each way of naming. It is in concrete toponymic practices that we best grasp relationships to places, the spatial inscription of power relations and their possible polarizations. The ceaseless reconfiguration of the overall landscapes, disconcerting as it may seem, organically expresses the interplay of social forces. It is based on three major principles: spontaneity, mobility and heterogeneity.

---

6. Hence, the need for the researcher to hear the languages in question. In Douala, the prefixes *Log*, *Ndog* and *Bona* signal the historical segmentation of neighborhoods founded by Bakoko (Logpom, Logbaba, etc.), Bassa du Wouri (Ndogkoti, Ndogbong, etc.) or Douala (Bonapriso, Bonabéri, etc.) lineages (Amougou Mbarga 2013, pp. 197–198).

7. There are two main methods of incorporating foreigners in African cities – by adding neighborhoods, whose name and internal toponymy will then reflect the culture of the newcomer (Boutillier 1969), or by directing newcomers to host neighborhoods, whose toponymic customs they will have to adopt.

### 11.2.1. *Spontaneity*

The fixing of names through use, rather than by a centralizing authority, provides information on what makes place, and for whom. Without formal registration, names are only valid as long as they make sense to those who use them. They "take" when they find their place in a universe of meaning, and disappear or are replaced when this is no longer the case. Here, we have therefore a reflection, partial but precious, of lived experience and of its collective meanings. This also includes the creation of generic terms, such as *kebbe* in Mauritania, from the verb *kebb*, to throw, to signify slums and the feeling of being thrown there, dumped. Names given spontaneously draw from very different registers of the ideological repertoire of states. Some simply reflect the physical characteristics of the place; others refer to an event that took place there, or to the history of a particular group, its cultural codes or its aspirations. Still others, especially in urban areas, draw from African or international news the material for commentary on the living conditions of the inhabitants[8]. As a general rule, the nationalist register favored by states is absent from spontaneous place naming. There, it is not even disputed but, more often than not, ignored (see in this book the chapter devoted to the naming of places in informal districts of southern cities by Melissa Wanjiru).

### 11.2.2. *Mobility*

Prior to its enmeshment in rigid borders, constitutive of (post)colonial states and their principle of contiguity, the African space included wide interstices between social or political formations, with borders themselves moving. These interstitial spaces were open to large-scale phenomena that continually transformed the overall configurations according to different temporalities – slow in the case of itinerant

---

8. Many African cities have neighborhoods named by their residents after hot spots in the news: Vietnam City in Tokombéré; Biafra in Abidjan; Soweto Village, Bosnia and Kosovo in the Nairobi slums; Beirut and Sarajevo in Brazzaville during the 1997 civil war; Gulf District in Bamako; Gaza Strip on the outskirts of Zanzibar. These names communicate, if not a high level of violence, at least a hard-felt marginality. Wanjiru (2018, pp. 115–116) notes that as the year 2000 approached, young residents of Nairobi's Mathare slum renamed their elders' Kosovo Village for themselves as New Millennium Village, a more positive and hopeful name. The two names presently coexist. Popular humor also mobilizes names evoking wealth and comfort, as much to mockingly rename neighborhoods of the newly rich (Las Palmas in Nouakchott) as to self-mockingly rename poor neighborhoods: Dallas or Las Vegas in Abidjan, Sétif or Kinshasa (Cheikh-Taine 1998; Dorier-Apprill 1999; Bertrand 2001; Garakcheme 2011; Wanjiru 2018; Yermeche 2018; personal observations).

agriculture, or faster in the case of wars or of the *exit option* of dominated groups. With the displacement of their habitat and their economic activities, migrant groups also brought with them the systems of representation that linked them, including toponymy. More than the replication of names, such as the Darou Khoudosses mentioned at the beginning of the chapter or the innumerable Toubas or Ifes (Ben Arrous 2016), the interest for CT lies in the mobility of the very conceptions of place and the ways of naming it[9]. The actualization of this principle of mobility, in contexts less open to the resettlement of migrants, involves adaptation strategies to which CT would benefit from paying more attention. The toponymy of prestige (official, honorific) is not always accessible to groups perceived as non-native. It is often skipped in favor of other toponymic inscriptions, names on shops, for example, or inscriptions on public transport vehicles, which relate to daily practices and uses. Here, too, territorializations are played out that are variably peaceful or contested.

### 11.2.3. *Heterogeneity*

The co-presence of mutually irreducible toponymic systems largely results from the preceding characteristic, that is, from the geographic mobility of value systems and conceptions of what constitutes place. Heterogeneity is the most troublesome principle with regard to the standardizing intentions of the name-giving state (and its expected benefits: legibility and control of the territory). Today, it is widest in cities and around the sites that consume the most migrant labor, extractive industries or large infrastructure projects. An example from Dakar will allow us to grasp the articulation of the three characteristics. At the colonial creation of Dakar, the Lebou villages on the site were destroyed. They were rebuilt further away, in other locations, and moved again over successive evictions but retaining their toponymic forms (*mobility*). In the aftermath of the plague epidemic of 1914, the colonial authorities, anxious to push all the African populations out of the Plateau, found themselves forced to resort to the Lebou nomenclature (*heterogeneity*), rather than location by street names, to locate those to be expelled. They adopted the Lebou terminology, which therefore "took" (*spontaneity*). One lexical creation became commonplace along the way: the *peñc*, which initially designated an assembly of decision-makers, has come to designate, to the present day, the Lebou places in Dakar (*spontaneity* again) (Bigon and Ben Arrous 2019).

The diversity of relationships to places might lead CT to somewhat relativize the importance of (re)namings in the process of territorialization. A lot depends on who

9. This is not exceptional in light of Europe's toponymic expansion. The latter is not limited to the dissemination of names like New York, New Orleans or New Zealand. It covers a much deeper process of diffusion of categories, forms and patterns of naming places.

names what, and who uses which names. "The name is one of the attributes of the territory", note Giraut et al. (2008, p. 98): "it designates it, situates it and should even qualify or symbolize its substance, even its essence." A name, however, is not always an essential attribute, and sometimes, it is not even an attribute at all.

The Shangwe refusal to fit into the colonial ethnic cartography provides a well-documented counter-example (Worby 1994). On the maps of the former Southern Rhodesia, now Zimbabwe, as early as the 1890s appeared a Shangwe "area", sometimes delimited, sometimes not. The Shangwe group, which was supposed to identify with it, and therefore "claim" this area as constituting its "customary lands", constantly avoided being named as such by the colonial authorities. It thus escaped registration as an ethnic subject governed by pseudo-customs, synonymous with subjection plain and simple (Mamdani 1996). By refusing the Shangwe ethnonym (and resorting, until the present day, to broader forms of self-referencing within the Shona group, or within the clan order at a lower level), the Shangwe at the same time disqualified the use of this name as a toponym. The name gradually disappeared from maps before independence. The lived territory, unnamed, remained.

Also instructive is the Arabization of Berber toponymy. At a surface level, Berber or Amazigh names in Algeria, Morocco and Tunisia seem to have resisted several centuries of Latinization, Arabization, Francization and re-Arabization. A militant current is working to de-Arabize (re-Berberize) what had been Arabized (de-Francized) in the first decades of independence in these countries. At a deeper level: what do we (re)name in Berber? The French authorities originally named "communes", which always seemed too big or too small, before borrowing from Arabic the category "douar". This category, carried over to independence, is more suitable in scale, but does not capture the substance of a segmented territoriality, based on the interplay of flexible alliances and networked solidarities (Adel 2005; Blais 2014). Neither the basic units of these alliances, larger or smaller than the douar, nor the territories they form are named (Figure 11.5). More profoundly still, the socio-spatial terminologies which describe them – that is, the words of the territory, such as *arch, ferqa, çof, leff* and so many others – all derive from Arabic. We might see there a dilution, a retreat of Berberity on the capital question of the territory, of its words and its names. But we can also consider the renunciation of specific toponymic categories as "the price to be paid" to leave the essential "more or less intact": the territorial structures themselves and the mechanisms of inheritance, of property and land transmission (Manzano 2006, p. 196). In this sense, place naming is not always the symbolic arena of identity representation that CT presupposes – at least not directly.

**Figure 11.5.** *View of Ich Neka douar in the Middle Atlas, Morocco. The douar is a convenient toponymic category, making it possible to identify groups of dwellings of a smaller size than the commune. In Berber country, however, as here, the douar does not constitute a social or political unit. It is home to different instances of power, competing within the same douar, but linked to equivalent instances in other douars, whether neighboring or more distant, forming wide-scale networks. The result is forms of territoriality that are difficult to capture through cartography. Toponymy does not capture them either (photo: Rachid Outahar, July 2014)*

## 11.3. A problem of hegemony

By focusing too much on governments and their manipulation of place naming, we risk seeing societies as mere passive receivers – which they are obviously not. It is easy to verify this in Africa. Here, the plurality of systems of naming makes official toponymy appear as an unequally shared cultural form. This is enough to reduce its scope as a political technology. But the contribution of (re)namings to the establishment or consolidation of power structures is no more guaranteed elsewhere. It is a latent contribution that only becomes effective when the governed adhere to the ideological vision promoted by the rulers. This adherence is what Gramsci called consent. It constitutes the true measure of the political effectiveness of place naming.

Drawing on Gramsci and his theory of hegemony, Puzey and Vuolteenaho (2016) recall that power is never so strong as when it organically represents the whole of society, and its legitimizing ideologies penetrate into common sense. It is then fully hegemonic. Failing that, it remains a system of domination which lacks the consent of the dominated. In this case, ideology and its vehicles (including place naming) have only a limited scope. CT would have much to gain from developing Gramscian approaches, argue the two authors. It would then integrate into its analyses a variable that is often lacking: the reciprocal relationship between rulers and ruled – where the real scope of toponymic operations essentially comes into play.

We support the theoretical approach of Puzey and Vuolteenaho while also mobilizing the notion of hegemony in a broad and figurative sense, that of undivided domination. Formal toponymy is in this sense a hegemonic construction under certain circumstances, not under others. It exercises an unequally consensual domination over ways of thinking about space and naming places. In Africa, its hegemony seems complete among the intellectual and political elites. The need to establish place names, if possible to their advantage, is obvious to them. The adherence of elites to a set of norms and principles, explicit like the devolution of the power of naming to politico-bureaucratic institutions, or implicit like the political instrumentalization of this power, confirms the existence of a more or less universalized model. The hegemony of the model is, however, mitigated by informal toponymies and by the (counter-hegemonic) practices deriving from them.

This broader conception of hegemony comes to temper the excessive idealism of analyses that wish to see in (formal) toponymy an arena of symbolic struggles for a more inclusive society, for more social justice or for the reparation of injustices (restorative justice). It also allows a better appreciation of the concrete impacts of place naming on the structuring of space, on its representations and on the power relations that may be read there but which usually elude CT's analyses. We illustrate this with two examples below.

The first takes up the questions of addressing and "decolonizing urban landscapes", mentioned at the beginning of the chapter, by showing what links them within the same model with hegemonic pretensions. The erasure of colonial street names is less massive on the African scale than its over-representation in CT studies suggests (Bigon and Ben Arrous 2018). The focus of studies on this theme is in itself a sign of adherence to a normative conception of what *should* be – but is not – the toponymy of African cities. The interest of an approach refocused on the model itself, rather than on specific names, is to reveal low-noise counter-hegemonic practices. The indifference of urban residents to street names and the persistent prevalence of unnamed streets are part of a coherent counter-hegemony, in which one can also detect reciprocal determinations between rulers and ruled.

The second example concerns an African-American researcher's proposal to rename Africa (Imhotep 2010). As quirky as this project may seem, the idea of imposing a more "African" name onto Africa fits perfectly with the logic of the dominant model. It also illustrates its limits: existing toponymic practices, which make the project unachievable.

### 11.3.1. *Street naming, symbolic power, and urban landscapes*

The relationship between these three elements is a matter of focus. By zooming in on a city center or a residential district from colonial times, one identifies streets that have changed names since independence. Depending on the extent of the changes observed, it may seem legitimate to discuss a decolonization of the urban landscape (or to deplore its insufficiency). Looking at cities from a wider view, we notice the preponderance of streets that have never been named. In these larger urban landscapes, the symbolic power of street naming requires re-examination.

A photographer would say that the decolonization of landscapes is also a matter of sharpness of image, even in the city center. A slightly blurred focus can be useful to highlight visual components other than names. Here, we must imagine a myopic gaze that could not clearly read signs, but would recognize their colors and shapes. White letters on a blue background in the former French colonies, not quite the same blue (and often also black letters on a white background) in the former British colonies, proportions and material of the signs, preferred methods of mounting (on walls or on posts): these standardized codes were an integral part of the visual identity of colonial towns. Their renewal or alteration, varying according to local and national political histories, remains understudied. Yet it would be worthy of investigation. The colonial odonymic *dispositif* is also perpetuated in material forms, just as conspicuous as the names themselves.

Moreover, the renaming of streets remains an act that is too specific, in space and time, for appreciating the "power" of odonymy. Only those streets which had one can be rid of an old name. Sixty years after the independence of most African countries, cities have spread well beyond their colonial limits. The majority of African city-dwellers therefore live in neighborhoods whose streets were never given colonial names (and often have no names at all). Nothing here is re-nameable. Even within the old boundaries, not all streets were given names. Those of the so-called indigenous neighborhoods were sometimes numbered, like the Medina of Dakar where the numbering established a social, cultural and above all racial distance from the Plateau (Bigon 2016). The most common state of affairs, however, was a lack of addressing. Places that made sense for the colonized (street corners, intersections, sub-districts, etc.) were named informally. Their names emerged from use, not from a colonial directive. Through their informal character, they have escaped the struggles for symbolic control of formal namescapes. They still escape these struggles, even though their relative autonomy with regard to street naming does not exclude mutual influences.

Thus, the commemorative principle of street naming, unknown to vernacular toponymies, may give rise to original appropriations. In Douala, for example, the Roger Milla, Thomas Nkono or Samuel Eto'o Fils crossroads do not celebrate political figures but the international glories of Cameroonian football (Amougou Mbarga 2013, p. 200). Conversely, administrative authorities are not completely unaware of informal names. "We are forced to use these 'vulgar names' to go into the field and even on the maps to properly [represent] the city", explains an addressing officer, again in Douala (op. cit., p. 201). As already mentioned, the colonial administration in Dakar had also found itself "forced", in order to identify (expel) the African populations of the Plateau, to use the Lebou toponymy rather than its own system of street naming. Urban transport operators, including public companies, also use informal place names: "The standard international practice is to name bus stops after streets, monuments or public establishments in structured areas. When the streets have no name or when you enter peri-urban areas, the general tendency in Africa is to choose names known to users, even if they are informal names"[10].

The uneven distribution of named streets in the urban space reduces residents' exposure to possible "messages" (political, ideological) from their rulers. On the other hand, it materializes the socio-spatial divisions resulting from colonial models of dual and segregated cities. These inequalities impose themselves on the inhabitants, independently of the ideologies proclaimed on street signs in the city center or in a few privileged neighborhoods. Maputo, the capital of Mozambique, is exemplary in this regard. In the aftermath of the war of independence (1964–1974), the authorities worked to inscribe a new symbolic order onto the odonymy of the town center. The names of Portuguese governors or generals were methodically replaced by names celebrating proletarian internationalism, African decolonizations and the Mozambican revolution. A network of numbered streets was created for outlying districts, halting at the gates of "irregular" neighborhoods (Figure 11.6). From these peripheral districts, the city center is perceived as an alien world. It is referred to as Xilonguine, "the place of the white people" in the Changana language,

10. Y.O., research officer at the African Union of Public Transport (UATP) in Abidjan, telephone interview, August 4, 2020.

and anyone who lives there, regardless of skin color, is called Mulungo ("White person")[11].

**Figure 11-6.** *An intra-urban boundary in Maputo, Mozambique. The street sign for Rua 3.468 marks an access point for the "City of Reeds" (Polana Caniço "A"), a material associated with the so-called indigenous neighborhoods of the former Lourenço Marques, the old colonial name of Maputo. It is the last element of signage before entering a neighborhood without addresses, but above all without basic infrastructure (water, sanitation, electricity), where living conditions are precarious. A few hundred meters away begins the "City of Cement" (Polana Cimento "A" and "B"), the other pole of colonial town planning, which now houses museums and ministries and whose street names reflect the highest revolutionary ideals. The unequal distribution of signage reproduces and materializes very wide economic and social disparities (photo: Torun Reite, September 2017)*

The spatial distribution of street names covers a geography of prestige, if not always of privilege. As a general rule, it concentrates upon the central or residential neighborhoods which colonial town planning itself had established as prestigious. In certain cases, this includes also more recent districts, ascribed to the middle or upper

11. Rural residents use the same term Xilonguine to designate Maputo as a whole, with no distinction between central and peripheral districts (Cumbe 2016, p. 198).

classes. By stratifying space and society, the distribution of street names constitutes a factor of *distinction*, in Bourdieu's sense: "The culture [which] contributes to the real integration of the dominant class", in this case the political culture which governs the choice of street names, "is also the culture which separates (the instrument of distinction) and which legitimizes distinctions by forcing all other cultures (designated as subcultures) to define themselves by their distance from the dominant culture" (1977, p. 408). Toponymies in common use are in fact designated by a characteristic of deviation: unofficial, informal, non-compliant. This characteristic takes precedence over other possible categorizations. Moreover, it blurs the boundaries between "vernacular", "traditional", "popular" and "alternative" toponymies – all affected, to varying degrees, by their common rejection as informal, and the constant reinventions resulting from it.

This form of domination is very far from a hegemony that might be qualified as "integral" in Gramsci's sense, in which odonymy would be "organically representative of society" (Puzey and Vuolteenaho 2016, p. 68). Other possibilities of interpretation arise when we consider the marginal character of street naming in the practice of urban residents. Contrary to the usual focus on operations of renaming, the lack of eagerness of certain municipalities to (re)name streets could be an indication of a closer organic relationship between rulers and ruled. In this hypothesis, the sovereign inaction of the municipalities would overlap with the indifference of the governed, reinforce it, perhaps even result from it.

The World Bank's addressing program, already mentioned, left to the authorities of some 50 cities the choice of naming or numbering streets. A few major axes have received names, but numbers prevailed everywhere[12]. As a member of the Dakar addressing commission explained: "If you give names, there will always be discontented people. A number is simpler, it means nothing, no one is going to pick a fight over nothing"[13]. The program concluded with a "sparing use of street signs", half as dense, on average, as what the World Bank had planned (Farvacque-Vitkovic et al. 2005, p. 50). This setback is modestly attributed to "budgetary constraints" (op. cit.) but lack of enthusiasm from the municipalities also played a role. It related, at least in part, to an elementary knowledge of the social environment: "We

12. The program covered the following countries between 1989 and 2004: Benin (3 cities), Burkina Faso (2), Cameroon (2), Congo (1), Côte d'Ivoire (3), Djibouti (1), Guinea (1), Mali (6), Mauritania (11), Mozambique (6), Niger (2), Rwanda (1), Senegal (11), Chad (1) and Togo (1).

13. B.T., interviewed on January 23, 2018. Similar comments were made by other members of addressing commissions in Niamey (Niger), September 7, 2018, and Conakry (Guinea), January 21, 2020.

suspected that the addressing would not 'take'. Whether names or numbers, the populations do not use them, it remains for them a 'government thing'"[14].

Pushing the interpretation a little further, we might consider the instrumentalization of street names as a foreign, imperial "language", an administrative and political language whose mastery was imposed by colonial history, but which remained the prerogative of elites and did not allow communication with the great majority. Indifference to names or numbers would then appear as a form of "decolonization of the mind", to provocatively borrow the title of a classic[15], reducing the role of this language in the writing of the city. Its supposed universality, in other words its claim to hegemony, would come up against both the counter-hegemonic practices of the governed (who name places in other ways) and, in some cases, the nonchalance of the rulers.

Whether they have changed a great deal or little since independence, street names only provide information about a *projected* symbolic order, not its reality. The latter closely depends on the reception of names, that is, the way in which they are read, ignored, collectively interpreted or reinterpreted. Nairobi, the Kenyan capital, is along with Maputo one of the cities whose odonymy has been the most updated. Names linked to British colonization were replaced under the presidency of Jomo Kenyatta by names that celebrated his own political party, KANU (the sole legal party), neglecting other currents of Kenyan society. This policy was intended to impose a monolithic narrative of the struggle for independence. In particular, it "suppressed" the memory of the Mau Mau movement, against which Kenyatta had constructed his own rise to power (Wanjiru 2018, pp. 75–78). The forbidden memory has not, however, disappeared from informal toponymy, as indicated by the Mau Mau Road which crosses the slum of Mathare (op. cit., p. 114), nor, more fundamentally, from the consciousness of several generations of Kenyans. On this level, the symbolic order projected by Kenyatta has not been realized.

At the other end of the spectrum, the Plateau in Dakar retains just over half of its old colonial street names. Several of them glorified the military conquest of Senegal. There remains the Rue Sandiniéry and the Rue de Thiong, which commemorated the occupation of the south of the country beginning in the 1860s. A semantic drift has occurred. When passers-by are asked about these names, they invariably reply that

14. B.T., interview cited above.

15. *Decolonizing the Mind* is a collection of essays by the Kenyan writer Ngũgĩ wa Thiong'o (1987) which argues for African writers to express themselves in their mother tongue rather than in European languages. Ngũgĩ's early novels and plays were in English; since then, he has written his works of fiction in Kikuyu (while retaining English for non-fiction). The book has become a classic of postcolonial studies.

they were given at independence to "teach about the geography of Senegal"[16]. The most imaginative argue that it was a matter of "pleasing the Casamançais" (i.e. the inhabitants of the south). But the most significant shift concerns the Avenue de la République, which unambiguously referred to the French Third Republic at the time of its colonial creation. Today, it is said to designate the republican form of independent Senegal. Here is another possible contribution to CT: the collective reinterpretation of old street names is no less rich in new meanings than the processes of renaming or neotoponymy. They still need to be studied over adequately long periods of time.

### 11.3.2. *Renaming Africa: a radical project and its limits*

Among the possible origins of the name Africa[17], Asar Imhotep cites the defeat of Hannibal by the Roman armies (Imhotep 2010, p. 5). The Romans used to name their conquered territories after the name of the last people militarily defeated. The Berber tribe Aourigha (whose pronunciation might perhaps have been Avrigha) would thus have given its name to the Roman province of Ifriqiya, corresponding more or less to present-day Tunisia. Such a name was not a tribute from the victors to the vanquished, but, on the contrary, a "mockery", an "insult", a form of humiliation reducing the inhabitants of the province to their sole quality of "defeated people" (op. cit., pp. 5–6). By giving herself another name, Africa and by extension the Africans of today would perform an act of self-determination (op. cit., p. 1). They would assert their right to define for themselves who they are and what they want to be (op. cit., p. 17).

At the *Deuxième Festival mondial des arts nègres*, held in Lagos in 1977, Wole Soyinka had already asked for the name Africa to be abandoned. His plea echoed Richard B. Moore's famous saying (1960, p. 49): "Slaves and dogs are named by their masters, free men name themselves"[18]. Soyinka argued that a new name should be rooted in an indigenous language. He proposed Abibirim or Abibiman, meaning in Akan "Black Earth" or "Land of the Blacks". But Imhotep considers these names to be too "reactive". He criticizes them for falling in line with an exogenous racial

16. Personal observations, January to February 2018. The two streets are well known to Dakar residents, one as an open-air fruit market and the other housing a barracks of the gendarmerie.

17. Zeleza (2004) lists seven competing etymologies, drawn from Berber, Latin, Greek, Phoenician, Hindi, the biblical Abraham's grandson Epher and the Yemenite king Africus.

18. Richard Benjamin Moore (1893–1978), who was born and died in Barbados, was an influential voice in the *Harlem Renaissance* movement. A bibliophile and self-taught historian, he was one of the first advocates of the term "African-American" to replace "Negro" or "Black".

thought, enclosing Africa in a skin color that is only black because "white Europeans" perceived it as such, historically, during their own self-definition as whites. These names would amount to (re)defining Africa as a double negative of Europe, that is, by what it is not (Imhotep 2010, pp. 9–10).

On the contrary, the new toponym should signify a deep identity, "serve a purpose" and indicate a way forward (op. cit., pp. 2, 6, 12). Based on these criteria, Imhotep opts for a programmatic name, in the manner of certain toponyms of pharaonic Egypt (Somaglino 2017). He proposes Dya Malela, which would mean "land of righteousness, justice, truth" and would also connote "love, friendship and fraternity" (Imhotep 2010, p. 13). The term would be the transcription in ciLuba (Thsiluba or Luba-Kasai) of an ancient name of Egypt. In hieroglyphic writing, its full meaning would be "a place cultivated (ripe, prepared) so one can expand (financially, intellectually, etc.), grow and mature (physically, spiritually, etc.) in his season within the context of a (strong, loving) community" (op. cit., p. 16).

Here, we have a seemingly radical project, aimed at renaming Africa – and therefore hundreds of millions of Africans – to "repair" the symbolic damage caused by imported toponymy. The approach is exemplary of an abstract restorative logic, disconnected from the human realities which it intends to influence. It is emblematic of a literalism very present in African studies of CT, where place names tend to be read according to the intentions of the (re)namers without reference to their re-readings by the societies concerned[19]. We might indeed doubt that many Africans are ready to abandon the name Africa on the grounds of a defeat of ancient Berbers by the Roman armies. The memory of this "insult", assuming Imhotep's interpretation to be correct, has had plenty of time to dissipate in favor of other meanings, consensual or disputed, multiple and changing depending on context, collective histories and individual perspectives.

A similar kind of abstract radicalism runs through the Facebook group *Renaming African Landmarks*. The group campaigns to restore the "original names" of the main African sites and monuments or, failing that, to give them new ones to replace

19. The importance given by CT to intentions is very reminiscent of the insistence in onomastics on the original meaning of place names. Onomastics relies on etymology to elucidate what names "mean" (or "meant" originally). CT favors the analysis of political intentions, that is, the symbolic order that the place-naming authorities would want to establish. Despite the somewhat caricatured accounts of the emergence of CT outside the field of onomastics, it is still a question of original meanings. Knowledge of these meanings is obviously useful, provided that a clear distinction is made between political intentions and their uncertain realization, inevitably conditioned by the interplay of social forces, confrontations and compromises.

colonial toponyms. Its flagship proposal is to restore the name of Victoria Falls to Mosi-oa-Tunya ("the smoke that thunders", in Silozi). However, this name never went away. It remains in use by Silozi-speaking local residents, as well as other names for speakers of other languages. The same applies to Lake Victoria. The group would like to rename it, regardless of the names it already has: Nyanza in Burundi, Ukerewe in Tanzania, Nalubaale in Uganda. The most paradoxical proposal is to rename Lake Victoria to Lake Thomas Sankara, in homage to the Burkinabè president who renamed Upper Volta as Burkina Faso or "land of upright men". As nice as it is, the idea of replacing a queen of England with a great man from Africa means adhering to a commemorative principle of colonial origin, fundamentally foreign to vernacular toponymies. Above all, Sankara's linguistic approach was of a different quality from what this renaming campaign is suggesting. His concern was to prevent a single ethnic group from using toponymy to elevate its language, and elevate itself, above others. The renaming of the country and its inhabitants encapsulated this, as "Burkina" is taken from the Mossi language and "Faso" from the Dyula language. Citizens from Burkina Faso are called Burkinabés, the -bè suffix coming from the Fulani language. This linguistic construction was explicitly designed to oppose the monopolization of the country's name by one dominating ethnic group.

The proposals of the Facebook campaign, like those of Imhotep or Soyinka, in each case favor *one particular* language, upon which they objectively confer a particular status. It might just as well be another language. In doing so, they simply transfer to the selected language the ascendancy that was once vested in the colonizer's language, without changing anything in the structure of domination or in the field of its inevitable contestation. The relationship between languages and hegemony has generally been established, in the margins of colonial toponymy, as one of opposition. Multilingualism is an important source of multi-toponymy. As such, it constitutes a non-negligible political resource, making it possible to thwart or at least reduce the symbolic domination of the (re)naming authorities.

It will be noted that the proposals discussed here pursue more or less the same objective. All aim to undo the grip of an exogenous toponymy, in order to find more organic links between places, their names, and people. But all make the attempt from an overhanging perspective, superbly ignoring the depth of the existing toponymic landscapes. The idea is always that the instrument of domination, if used with different intentions, might serve as an instrument of emancipation. It would be up to rulers or moral authorities (Soyinka has this status, Imhotep aspires to it) to instill a virtuous dynamic. This leads to an overestimation of both the penetration of toponymy of colonial origin into societies (which one would wish to emancipate) and the political effectiveness of this inherited model. Consent to the hegemonic claims of the dominant model conflicts with real practices, in particular the (always provisional) fixation of names through practice and the volatility of their meanings. Ndlovu-Gatsheni (2013) notes that a "principle of coloniality" crept into the institutions of postcolonial states, more or less "modeled" according to the standards of the former metropolises. It also seems to have crept into the toponymic means of envisaging symbolic emancipation.

## 11.4. Making sense: a heuristic of practices

How are place names received, perceived, used? Which ones are employed under what circumstances? What values do they hold depending on the use we make of them? How, and through what practices, are new names integrated into daily routines or rejected? What do their material inscriptions in space signify, and for whom? How do they relate to other names that are not materialized through signage? How does each participate in an overall sociability, its reproduction or its transformations?

"It is by our use of things, and what we say, think and feel about them – how we represent them – that we give them a meaning", taught Stuart Hall (1997, p. 3). The cultural theorist invited his peers to seek the meaning of "things" (i.e. material objects as well as ideas and cultural facts) in "meaningful practices". This seems to us to be sound methodological advice for CT. Clearly, in Africa, practices give meanings to formal toponymy that are sometimes far removed from what the naming institutions would have wished – and this is no doubt the case elsewhere as well.

Appeals to "explore the political economy of toponymic practices" are not new (Rose-Redwood et al. 2010, p. 453). However, the practices that come under the most scrutiny remain the official practices of naming (Giraut and Houssay-Holzschuch 2016). In comparison, issues of reception receive marginal attention (Myers 1996). Their aspect that has been most studied (or the least understudied) is the acceptance, or lack thereof, of name changes: how to explain why new toponyms do not "take"? This question might be addressed in the most diverse contexts, African or otherwise. However, CT tends to reserve it for situations of regime change, where old heroes are de-commemorated so as to commemorate new ones. In these very particular contexts, the non-appropriation of neo-toponyms is easily perceived through the prism of a more or less deliberate "resistance", of conscious and militant opposition to a new ideological order. "Resistance" is indeed an "important practice", the authors of a study on Bucharest concede, but it "does not explain every instance where place names fail to find popular acceptance" (Light and Young 2014, p. 672). Another study, on Timisoara, notes that the new leaders lacked a pedagogic approach, chose "martyr names" that were ill-suited to the "specificity of space", and neglected "practical aspects" by introducing toponyms that were too long and difficult to memorize (Creţan and Matthews 2016, p. 101).

Other explanations remain to be devised for more ordinary situations, when neither the political order nor the ideological order have been overturned but, despite everything, neotoponyms have not found takers. The question of non-appropriation is worth extending to already existing names, or even, in African situations, to little-used categories such as street names. Naming policies, their context and the names themselves usually communicate the intentions of the namers without too much difficulty. Nothing so transparent makes it possible to probe the motivation – hypothetical, conjectured – of those who do not adopt certain names. But perhaps this is not necessary. When names are neither used nor disputed nor even commented on, a finding of indifference is sufficient. The governed have other toponymic resources at their disposal than official nomenclature, other ways of naming places. These resources are of course more abundant when several systems of naming coexist, as is the case in Africa. But any change of name, in the former Eastern Bloc or elsewhere, actually creates a situation of multi-toponymy, at least temporarily. This lasts until the new official names replace, in ordinary practice, old names that have become "informal". It continues if they fail to do so. CT's natural inclination is to contrast the intentions of those in power to contrary intentions, that is, to a "resistance" conjectured in a more or less speculative way. It seems more judicious, on a methodological level, to contrast these same intentions to the practices, tangible and observable, of everyday life.

Multilingual toponymies are another instance of multi-toponymy. The "material turn" in the social sciences (Mukerji 2015) has created bridges between CT and sociolinguistics. The study of "linguistic landscapes", that is, of the physical inscription of languages and their visibility in public space, proves useful for identifying gaps, sometimes significant ones, between official language policies and their concrete expression. Deliberately or not, one or more languages are always more prominent than others on public buildings, commercial signs, or billboards. Signage of place names is no exception. As such, it constitutes a target of choice for radical groups, whether hostile to state monolingualism or, on the contrary, to an already established multilingualism. The recognition of Tamazight (Berber) as a national and official language in Algeria came after a virulent campaign known in the mid-2010s as the "de-Arabization of Kabyle toponymy", where Arabic inscriptions on signs were selectively covered with black paint (Sabri 2019)[20].

20. The selective damage of multilingual inscriptions is a worldwide phenomenon. In September 2020, a Google search with the terms "bilingual signage" and "vandalization" generated 1,700 results ranging from Calgary to Jerusalem and from Bayonne to Vukovar.

**Figure 11.7.** *Seat of the Communal People's Assembly (APC) of Tichy in Kabylia, Algeria. On the black wall in the foreground, erected before independence, the inscription in Arabic remains subordinate, crushed by the size of the inscription in French. The red and green panels on the left of the photo are newer. With characters of equal size, they now place Arabic at the top of the hierarchy ("official state language"), followed by Berber ("official language") and French. Although this latter is now a "foreign" language, it remains admissible on the pediment of an official building (photo: Habib Kaki, February 2016)*

But more peaceful linguistic landscapes also provide valuable insights into the reciprocal relationship between rulers and ruled and how it is negotiated. Figure 11.7, again in Kabylie, sheds a nuanced light on the "war of languages" in Algeria[21]. The photo shows two generations of toponymic inscriptions: one, colonial, is in French and Arabic (which was at the time a "foreign language") and

21. For the record, the colonial authorities declared Arabic to be a "foreign language" in 1938 (Gafaïti 2002, p. 23). At independence in 1962, it became the only official language. A policy of Arabization of place names was enacted, but classical Arabic was used rather than Darija (the colloquial Arabic in popular use), hindering the acceptance of new names. Breaking with this policy, Berber cultural claims were stifled for several decades, repressed, condemned in the state media as an "instrument in the service of neocolonialism". Berber was recognized as a "national language" in 2002, then as a "national and official language" in 2016. However, it remains at a level below Arabic, the only "official state language" in the constitution.

the other is in Arabic, Berber and French again (although this language in turn became "foreign" at independence). The inscription "Ville de Tichy", a relic of the colonial administration, has remained in place. Note the maintenance of the final *y*, Frenchified in the manner of Vichy or Clichy, while the transcription of the Berber toponym would rather be Ticci. We have here a set of compromises, adjustments of signage to real linguistic uses, which contrast with the rigidity of officially proclaimed linguistic and toponymic policies.

**Figure 11.8.** *Illegible signage in Lubumbashi, DRC. The wear of time, bad weather and lack of maintenance have erased the text from the signs. Their state of neglect contrasts with the visibly meticulous care given to expensive cars (photo: Donatien Dibwe dia Wembu, October 2017)*

The boundaries of the linguistic landscape are, for CT, the writing of place names. Potentially significant visual elements are, however, not all written. To integrate non-verbal elements into the analysis, Jaworski and Thurlow (2010) suggested the broader notion of "semantic landscape". The built environment constitutes, for example, an element of semantic landscape when its intra-urban variations – architecture, materials and forms of buildings, the width of roads, street furniture, public lighting or its absence, etc. – are superimposed onto the inequalities of odonymic treatment – streets named, numbered, or neither. Other elements of semantic landscape may include the visual codes of colonial signage (already mentioned), their degree of renewal, the maintenance of the signs, their wear or their replacement, the materiality of the makeshift inscriptions which repeat them, comment on them or cover them up (Fraenkel et al. 2015; Bigon and Ben Arrous 2018). "Semantic" or "linguistic", significant landscapes for CT are those that show not only the spatial inscription of toponyms but also their uses and their non-uses. An illegible inscription (Figure 11.8) or an absent address (Figure 11.9) may be rich in meaning. Stuart Hall would say that the non-use of things, as much as their use, can be a meaningful practice.

**Figure 11.9.** *Front of a carpentry shop on the Avenue du Château d'eau, Niamey, Niger. The sign mentions two different methods of locating the carpentry shop according to landmarks, but not the avenue where it is located. Other information is provided, allowing the company to be contacted (four telephone numbers, a post office box number, an email address) and for verifying its formality (tax identification and commercial register numbers). The non-mention of the name of the avenue suggests that it is not considered useful data (photo: Michel Ben Arrous, September 2018)*

Visual approaches also have their limits. Informal place names may not always be visible. Not all practices are material. Relationships between toponymy, practices and meanings shape individual spaces of freedom that escape the eye of the observer as much as that of states. It is therefore important to listen to what actors say about their own practices, including the most routine acts such as finding one's way or moving around: "Traditionally my neighborhood is called Ndoghem", explains a resident of Douala: "Normally we call it Cité-Sic and sometimes Ange Raphaël. In fact, depending on the time of day and where I am, I choose the appropriate name to facilitate my transport by taxi" (Amougou Mbarga 2013, p. 203). The observer might be disappointed to find nothing spectacularly political in this strategy – common to all the big cities of Africa – which consists simply of using the names that are the best for communicating with the interlocutor. Such a strategy, however, testifies, by its very banality, to the capacity of individuals (and societies) to evade the imposition of a single toponymy and its possible instrumentalizations.

We find this faculty of evasion in Algiers, if we ask the younger generations about their a priori sacrilegious toponymic choices. Young Algérois use the old names of streets and neighborhoods, often of colonial origin, rather than the official names commemorating the *shuhada* (martyrs) and *mujahideen* (combatants) of the war of independence. A hurried observer might conclude a lack of historical or political awareness. But those involved explain that the new names suggest a model of war heroes in which they do not recognize themselves. Their political imagination is no longer that of their elders (Bedjaoui 2017). In escaping the grip of official toponymy, they withdraw their consent for its instrumentalization. In doing so, they neither reject nor promote any particular ideology, but delegitimize the marking of their daily space by any form of state ideology.

"Our studies of technologies of power have often led us to believe their own statements of efficacy," notes Michael Crang (2000, p. 137) concerning, among other things, manipulations of toponymy and a certain naivety of his fellow cultural geographers. The plasticity of practices and the instability of meanings confirm that in contemporary Africa, the instrumentalization of toponymy suffers from a lack of efficacy and sometimes even, as with young Algérois, from a lack of legitimacy. CT cannot discern these dynamic aspects based on official operations of (re)naming alone. It is in the bustle of daily life, in its small "quiet geographies" (Di Méo 1999), that we best perceive the *relationships* between co-present toponymic systems and their social and political meanings.

## 11.5. Final remarks

This chapter has shown the limits, in African situations, of toponymic analyses that focus on official (re)naming. Contrary to the common Africanist perspective, which interprets its objects according to pre-established theoretical models, we have proposed a heuristic reversal. This would consist of relying on the diversity of forms of place naming, on the uses that societies make of them and the practices that give them meaning, to renew the objects, methods, issues and conceptual horizons of CT. We suppose that such a renewal would be useful for the interpretation of other toponymic landscapes, not just African ones. It will be for other studies, on other sites, to verify this.

A CT that is more attentive to the lack of hegemony of formal toponymy would, in turn, contribute to overcoming the nationalist paradigm that plagues political intelligence in contemporary Africa and, more widely, in postcolonial situations. Before-and-after comparisons of street names, centered on national independences, suffer from the same impoverished vision as the commonplace three-stage periodizations (precolonial, colonial, postcolonial). They hide more than they reveal. By reducing the history of ex-colonized societies to a linear march from domination to sovereignty, they obscure everything that cannot be reduced to it. In particular, they erase the historicity of the "subaltern" groups of yesterday or today, to borrow from Indian *Subaltern Studies* the term which they borrow from Gramsci, or "social cadets" as is more often said in Africa (Diouf 1999; Del Roio 2007; Liguori 2016)[22]. They also overlook the manufacture of consent to relatively stabilized forms of domination over the long term (Bin Wong 2003). These missing, obliterated stories, however, constitute the background to collective mobilizations in the present day. In situations of multi-toponymy, the names that are actually used in practice, whether by the greatest number or by singular groups, make it possible to restore this background, at least in part. Concrete uses of toponymy shed light on the state of consensuses, their topicality and their possible transformations.

From the beginning of this chapter, we have encountered spontaneous forms of toponymy, through which current social formations are embedded in disparate historical durations. Reference to kingdoms issuing from the Djolof empire continues to make sense for the Saloum-Saloum or the Baol-Baol of today, after

22. In his *Prison Notebooks*, written between 1929 and 1935, Gramsci refers to as "subaltern" both the "most marginal" social groups, characterized by the absence of "class consciousness", and those classes not integrated into the "dominant bloc" which decides the political–historical direction of the state. For Gramsci, emancipation is only possible *by* and *within* the state. Indian *Subaltern Studies* have freely minimized the question of the state in order to privilege the relative autonomy of subaltern subjectivities and to document, in particular, revolts by the Indian peasantry against the colonial authority. By transposing this reformulation to the field of CT, we note that the focus on official toponymy as an arena of symbolic struggle is part of the Gramscian vision of "no salvation outside the state". Multi-toponymy, on the contrary, involves disparate registers of references, relatively independent of the major narratives of the dominant bloc.

having served as a spatial framework for the economic and social recompositions of the 20th century – groundnut specialization and the influence of Sufi Islam. The same goes for Fuladu, which had only a few decades of autonomous existence before European colonization. The advance of the groundnut frontier (and of Sufi brotherhoods) in Fuladu polarizes collective actors on the basis of heterogeneous temporalities, which overlap more than they succeed one another. It is difficult to analyze this situation (or others, similarly inscribed in multiple temporalities) by seeing the colonial moment as a circumscribed and closed period. From a dual historiographical and political point of view, it is more judicious to consider it as a "great bifurcation"[23], which to a greater or lesser extent diverted the social formations from the trajectories they followed up until then – but did not necessarily stop them or reduce them, after independence, to following a single national trajectory.

Over shorter periods of time, the colonial powers introduced their own toponymic forms on a wide scale – directly, in the case of street naming, or in a more indirect way, with regard to the ethno-linguistic classifications upon which the spatial inscriptions of ethnicity and the claims of autochthony that accompany them are based today. The way these forms "take", or not, varies according to place, context and particular political histories. These variations are sufficient to indicate the circumstantial character of the reappropriations, that is, once again, a set of selective mobilizations referring to multiple and entangled historicities.

Conventional CT strives to perceive toponymic landscapes as "discursive formations" (Giraut and Houssay-Holzschuch 2016, p. 5) but with a single speaker – a kind of monologue in which (post)colonial rulers alone would construct or consolidate "power structures through linguistic means" (Puzey and Vuolteenaho 2016, p. 66). Multi-toponymy and its uses invite us to pass from the discursive principle to another that we might almost – but not quite – qualify as "dialogical", after Bakhtin. Dialogical, in the sense that each discursive statement, each inscription or toponymic stratum, always constitutes a "response" to the statements preceding it and determines, to varying degrees, the meaning of those that come after (Todorov 1981). But not quite, because we are dealing here with a set of simultaneous "dialogues", whereby each element of a toponymic landscape simultaneously takes on different meanings in relation to several other elements. It is not always easy to identify which power relations are formed or reconfigured in this complex game of utterances and interactions. But it is only thus that we can perceive

23. We borrow the expression from Jean-François Bayart (2016) who borrows it from Prasenjit Duara (1995).

the subaltern voices which the nationalist paradigm ordinarily stifles. A renewed CT might also be useful in this regard.

## 11.6. References

Adebanwi, W. (2017). Coloring rainbow streets: The struggle for toponymic multiracialism in urban post-apartheid South Africa. In *The Political Life of Urban Streetscapes: Naming, Politics, and Place*, Rose-Redwood, R., Alderman, D., Azaryahu, M. (eds). Routledge, New York, 218–239.

Adel, K. (2005). Dans l'Aurès : de l'Ahmar Khaddou à M'zirâa. Changements et permanences. In *L'espace montagnard entre mutations et permanences*, Messaci-Belhocine, N. (ed.). CRASC, Oran, 141–171.

Amselle, J.-L. (1985). Ethnies et espaces : pour une anthropologie topologique. In *Au cœur de l'ethnie. Ethnies, tribalisme et État en Afrique*, Amselle, J.-L. and M'Bokolo, E. (eds). La Découverte, Paris, 11–48.

Amougou Mbarga, A.B. (2013). À travers les dénominations des rues et des quartiers de la ville de Douala : la quotidienneté comme univers de sens. *Anthropologie et Sociétés*, 37(1), 195–212.

Anyinam, C. (1999). Ethnomedicine, sacred spaces, and ecosystem preservation and conservation in Africa. In *Sacred Spaces and Public Quarrels: African Cultural and Economic Landscapes*, Kalipeni, E. and Zeleza, P.T. (eds). Africa World Press, Trenton and Asmara, 127–146.

Augé, M. (1992). *Non-lieux. Introduction à une anthropologie de la surmodernité*. Le Seuil, Paris.

Bates, R.H., Mudimbe, V.Y., O'Barr, J. (ed.) (1993). *Africa and the Disciplines: The Contributions of Research in Africa to the Social Sciences and Humanities*. University of Chicago Press, Chicago.

Batoma, A. (2006). African ethnonyms and toponyms: An annotated bibliography. *Electronic Journal of Africana Bibliography*, 10(1) [Online]. Available at: https://doi.org/10.17077/1092-9576.1000 [Accessed 13 March 2020].

Bayart, J.-F. (1996). *L'illusion identitaire*. Fayard, Paris.

Bayart, J.-F. (2016). Une critique politique de la mémoire : leçons d'Afrique (et d'ailleurs) [Online]. Available at: https://blogs.mediapart.fr/jean-francois-bayart/blog/130716/une-critique-politique-de-la-memoire-lecons-d-afrique-et-d-ailleurs [Accessed 13 June 2020].

Bedjaoui, W. (2017). La toponymie à Alger : les décisions glottopolitiques tiennent-elles compte des usages réels ? Étude de sociolinguistique urbaine. *Socles*, 10, 17–30.

Ben Arrous, M. (2009). La géographie par le bas : introduction à une aventure collective. In *Etudes africaines de géographie par le bas*, Ben Arrous, M. (ed.). CODESRIA, Dakar, 2–38.

Ben Arrous, M. (2016). Sarah's globe and the (un-)naming of mobile space. In *Place Names in Africa: Colonial Urban Legacies, Entangled Histories*, Bigon, L. (ed.). Springer, Cham, 27–36.

Bernus, E. (1980). L'arbre dans le nomad's land. *Cahiers de l'ORSTOM (série Sciences Humaines)*, XVII(3–4), 171–176.

Bertrand, M. (2001). Dynamiques urbaines, composition toponymique : le cas de Bamako (Mali). In *Nommer les nouveaux territoires urbains*, Rivière d'Arc, H. (ed.). Editions de la Maison des sciences de l'homme, Paris, 250–272.

Bigon, L. (2016). *French Colonial Dakar. The Morphogenesis of an African Capital.* Manchester University Press, Manchester.

Bigon, L. and Ben Arrous, M. (2018). Street signage: From here to urban Africa and back again. Exhibition catalogue, Beit Ha'Adrichal, Jaffa.

Bigon, L. and Ben Arrous, M. (2019). Les noms de rues à Dakar : héritages (pré-)coloniaux et temps présent. *Bulletin de l'IFAN (Série B: Sciences Humaines)*, LIX(1–2), 53–75.

Bin Wong, R. (2003). Between nation and world: Braudelian regions in Asia. *Review – A Journal of the Fernand Braudel Center*, 26(1), 1–45.

Blais, H. (2014). *Mirages de la carte. L'invention de l'Algérie coloniale.* Fayard, Paris.

Bourdieu, P. (1977). Sur le pouvoir symbolique. *Annales : économies, sociétés, civilisations*, 32(3), 405–411.

Boutillier, J.-L. (1969). La ville de Bouna : de l'époque précoloniale à aujourd'hui. *Cahiers de l'ORSTOM (série Sciences Humaines)*, VI(2), 3–20.

Chaudhuri, A. (2004). In the waiting-room of history. *London Review of Books*, 26(12) [Online]. Available at: https://www.lrb.co.uk/the-paper/v26/n12/amit-chaudhuri/in-the-waiting-room-of-history [Accessed 12 June 2020].

Cheikh-Taine, C. (1998). Toponymie et urbanisation. In *Espaces et sociétés en Mauritanie*, Frérot, A.-M. (ed.). Urbama, Tours, 77–86.

Costa Dias, E. (2006). Regulado do Gabú (1900–1930): A difícil compatibilização entre legitimidades tradicionais e a reorganização do espaço colonial. *Africana Studia*, 9, 99–125.

Crang, M. (2000). Relics, places and unwritten geographies in the work of Michel de Certeau (1925–86). In *Thinking Space*, Crang, M. and Thrift, N. (eds). Routledge, London, 136–153.

Creţan, R. and Matthews, P.W. (2016). Popular responses to city-text changes: Street naming and the politics of practicality in a post-socialist martyr city. *Area*, 48(1), 92–102.

Cumbe, C. (2016). Formal and informal toponymic inscriptions in Maputo: Towards socio-linguistics and anthropology of street naming. In *Place Names in Africa*, Bigon, L. (ed.). Springer, Cham, 195–205.

Cumings, B. (1998). Boundary displacement: Area studies and international studies during and after the cold war. In *Universities and Empire: Money and Politics in the Social Sciences During the Cold War*, Simpson, C. (ed.). The New Press, New York, 159–188.

Daramy, A. (2019). The art of drawing your address in The Gambia [Online]. Available at: www.bbc.com/news/world-africa-47968968 [Accessed 21 July 2020].

Del Roio, M. (2007). Gramsci e a emancipação do subalterno. *Revista de Sociologia e Política*, 29 [Online]. Available at: https://www.scielo.br/pdf/rsocp/n29/a06n29.pdf [Accessed 10 April 2020].

Di Méo, G. (1999). Géographies tranquilles du quotidien : une analyse de la contribution des sciences sociales et de la géographie à l'étude des pratiques spatiales. *Cahiers de géographie du Québec*, 43(118), 75–93.

Diouf, M. (ed.) (1999). *L'historiographie indienne en débat. Colonialisme, nationalisme et sociétés postcoloniales*. Karthala and Sephis, Paris and Amsterdam.

Dorier-Apprill, E. (1999). Brazzaville : des quartiers pour territoire ? In *La nation et le territoire, vol. 2, Le territoire, lien ou frontière ?*, Bonnemaison, J., Cambrézy, L., Quinty-Bourgeois, L. (eds). L'Harmattan, Paris.

Dorier-Apprill, E. and Van den Avenne, C. (2002). Usages toponymiques et pratiques de l'espace urbain à Mopti (Mali). *Marges linguistiques*, 3, 152–158.

Duara, P. (1995). *Rescuing History from the Nation: Questioning Narratives of Modern China*. University of Chicago Press, Chicago.

Emballo, C. (2007). *Retratos da Casamance*. Novagrafica, Bissau.

Farvacque-Vitkovic, C., Godin, L., Leroux, H., Verdet, F., Chavez, R. (2005). *Adressage et gestion des villes*. World Bank, Washington, D.C.

Fraenkel, B., Cumbe, C., Cozzolino, F. (eds) (2015). *Escritas urbanas. Ler, escrever, agir na cidade*. Alcance, Maputo.

Gafaïti, H. (2002). The monotheism of the other. Language and de/construction of national identity in postcolonial Algeria. In *Algeria in Other's Languages*, Berger, A.-E. (ed.). Cornell University Press, Ithaca, 19–43.

Garakcheme, G. (2011). Références violentes et toponymie des quartiers à Tokombéré. *Sociétés et jeunesses en difficulté*, 11 [Online]. Available at: http://journals.openedition.org/sejed/7203 [Accessed 20 March 2020].

Geschiere, P. (1998). Le social standardisé : l'État contre la communauté ? Revue critique du dernier livre de James C. Scott, "Seeing Like a State – How Certain Schemes to Improve Human Condition Have Failed". *Critique internationale*, 1, 60–65.

Giraut, F. and Houssay-Holzschuch, M. (2016). Place naming as *Dispositif*: Toward a theoretical framework. *Geopolitics*, 21(1), 1–21.

Giraut, F., Houssay-Holzschuch, M., Guyot, S. (2008). Au nom des territoires : enjeux géographiques de la toponymie. *L'Espace géographique*, 37(2), 97–105.

Gramsci, A. (1996). *Cahiers de prison*, 5 vols. Gallimard, Paris.

Guyer, J. (1995). A perspective on African studies in the United States. Report, Ford Foundation, New York.

Hall, S. (1997). Introduction. In *Representation: Cultural Representations and Signifying Practices*, Hall, S. (ed.). Sage, London, 1–11.

Hountondji, P. (ed.) (1994). *Les savoirs endogènes : pistes pour une recherche*. CODESRIA, Dakar.

Imhotep, A. (2010). Renaming Africa [Online]. Available at: www.asarimhotep.com/documentdownloads/Renaming_Africa.pdf [Accessed 15 March 2020].

Jaworski, A. and Thurlow, C. (2010). Introducing semiotic landscapes. In *Semiotic Landscapes: Language, Image, Space*, Jaworski, A. and Thurlow, C. (ed.). Continuum, London, 1–40.

Kam Kah, H. (2019). Come-no-go/l'ennemi… dans la maison: Reflections on the Lingoes of conflict in Cameroon's urban history. *Modern Africa: Politics, History and Society*, 7(1), 137–160.

Light, D. and Young, C. (2014). Habit, memory and the persistence of socialist-era street names in post-socialist Bucharest, Romania. *Annals of the Association of American Geographers*, 104(3), 668–685.

Liguori, G. (2016). Le concept de subalterne chez Gramsci. *Mélanges de l'École française de Rome – Italie et Méditerranée modernes et contemporaines*, 128(2) [Online]. Available at: http://journals.openedition.org/mefrim/3002 [Accessed 8 February 2020].

Luig, U. and Von Oppen, A. (1997). Landscape in Africa, process and vision: An introductory essay. *Paideuma*, 43, 7–45.

Mamdani, M. (1996). *Citizen and Subject: Contemporary Africa and the Legacy of Late Colonialism*. Princeton University Press, Princeton.

Mamdani, M. (1997). Africa and "African Studies". In *Knowledge, Identity and Curriculum Transformation in Africa*, Cloete, N., Muller, J., Makgoba, M.W., Ekong, D. (eds). Maskew Miller Longman, Cape Town, 149–154.

Manzano, F. (2006). Berbères, berbérité : noms, territoires, identités. *Cahiers de sociolinguistique*, 11, 175–214.

Marshall-Fratani, R. (2006). The war of "who is who": Autochthony, nationalism, and citizenship in the Ivoirian crisis. *African Studies Review*, 49(2), 9–43.

Mavungu, E.M. (2016). Frontiers of power and prosperity: Explaining provincial boundary disputes in postapartheid South Africa. *African Studies Review*, 59(2), 183–208.

Mokwena, L. (2020). Blurring the boundaries between "urban" and "rural": Oral route directions and the invention of street names in Ulco and Delportshoop. *South African Geographical Journal* [Online]. Available at: https://doi.org/10.1080/03736245.2020.1786443 [Accessed 11 August 2020].

Moore, R.B. (1960). *The Name "Negro" – Its Origin and Evil Use*. Afro-American Publishers, New York.

Mouiche, I. (2013). Dénomination et territorialité urbaines, chefferies traditionnelles et question identitaire en pays bamiléké au Cameroun. *Autrepart*, 64, 37–54.

Mukerji, C. (2015). The material turn. In *Emerging Trends in the Social and Behavorial Sciences*, Scott, R.A. and Kosslyn, S.M. (eds). Wiley Online Library [Online]. Available at: https://doi.org/10.1002/9781118900772.etrds0109 [Accessed 15 September 2020].

Myers, G.A. (1996). Naming and placing the other: Power and the urban landscape in Zanzibar. *Tijdschrift voor Economische en Sociale Geografie*, 87(3), 237–246.

Ndletyana, M. (2012). Changing place names in post-apartheid South Africa: Accounting for the unevenness. *Social Dynamics*, 38(1), 87–103.

Ndlovu-Gatsheni, S.J. (2013). *Coloniality of Power in Postcolonial Africa: Myths of Decolonization*. CODESRIA, Dakar.

Ndlovu-Gatsheni, S.J. (2018). *Epistemic Freedom in Africa: Deprovincialization and Decolonization*. Routledge, London.

Neotopo (2018). Academic bibliography on African place naming [Online]. Available at: https://neotopo.hypotheses.org/1191 [Accessed 13 March 2020].

N'Gaïdé, A. (2009). Identités ethniques et territorialisation en Casamance. In *Études africaines de géographie par le bas*, Ben Arrous, M. (ed.). CODESRIA, Dakar, 39–77.

N'Gaïdé, A. (2015). Logiques d'héritages et superposition de droits : conflits de pratiques dans l'Afrique contemporaine, le "légitime" contre le "légal" ? In *Land in the Struggles for Citizenship in Africa*, Moyo, S., Tsikata, D., Diop, Y. (eds). CODESRIA, Dakar, 69–97.

Ngũgĩ, W.T. (1987). *Decolonising the Mind: The Politics of Language in African Literature*. Heinemann, Nairobi.

Onguéné Essono, L.M. (2018). Dénomination nouvelle et abrogation mémorielle des lieux originels de Yaoundé. *Symposium Nommer les lieux en Afrique*, Université Abdou Moumouni and Université de Genève, Niamey.

Puzey, G. and Vuolteenaho, J. (2016). Developing a Gramscian approach to toponymy. In *Names and their Environment: Proceedings of the 25th International Congress of Onomastic Sciences*, vol. 2, Hough, C. and Izdebska, D. (eds). University of Glasgow, Glasgow, 66–77.

Rose-Redwood, R., Alderman, D., Azaryahu, M. (2010). Geographies of toponymic inscription: New directions in critical place-names studies. *Progress in Human Geography*, 34(4), 453–470.

Sabri, M. (2019). Le barbouillage des panneaux de signalisation en Kabylie : un moyen de revendication linguistique et identitaire. Working paper, Université Mouloud Mammeri de Tizi-Ouzou [Online]. Available at: https://dl.ummto.dz/handle/ummto/11439 [Accessed 17 September 2020].

Scott, J.C. (1998). *Seeing Like a State – How Certain Schemes to Improve Human Condition Have Failed.* Yale University Press, New Haven.

Sidibé, M. (2002). Entre le Saloum et la forêt de Pata : mobilité des migrants, intégration des espaces. *Revue européenne des migrations internationales*, 18(2), 189–207.

Sinou, A. (2014). The "plateau" in West African, French-speaking colonial towns: Between garden and city. In *Garden Cities and Colonial Planning. Transnationality and Urban Ideas in Africa and Palestine*, Bigon, L. and Katz, Y. (eds). Manchester University Press, Manchester, 74–97.

Somaglino, C. (2017). La toponymie égyptienne en territoires conquis : les noms-programmes des *menenou*. In *Du Sinaï au Soudan. Itinéraires d'une égyptologue (Mélanges offerts à Dominique Valbelle)*, Favry, N., Ragazzoli, C., Somaglino, C., Tallet, P. (eds). Éditions de Boccard, Paris, 229–242.

Soyinka, W. (1977). The scholar in African society. In *Second World Black and African Festival of the Arts and Culture: Colloquium on Black Civilization and Education. Colloquium Proceedings*, vol. 1, Iwara, A.U. and Mveng, E. (eds). Federal Government of Nigeria, Lagos, 44–53.

Todorov, T. (1981). *Mikhaïl Bakhtine. Le principe dialogique, suivi de Écrits du cercle de Bakhtine.* Le Seuil, Paris.

Topalov, C. (2002). Langage, société et divisions urbaines. In *Les divisions de la ville*, Topalov, C. (ed.). UNESCO and Maison des sciences de l'homme, Paris, 375–449.

UN-Habitat (2016). World cities report 2016. Report, United Nations Human Settlements Programme, Nairobi.

Von Oppen, A. (2003). Bounding villages. The enclosure of locality in Central Africa, 1890s to 1990s. HDR Thesis, Humboldt University, Berlin.

Vuolteenaho, J. and Berg, L.D. (2009). Towards critical toponymies. In *Critical Toponymies: The Contested Politics of Place Naming*, Berg, L.D. and Vuolteenaho, J. (eds). Ashgate, Farnham, 1–18.

Wade, S. (2009). Les associations urbaines et le pouvoir local dans les quartiers populaires. In *Etudes africaines de géographie par le bas*, Ben Arrous, M. (ed.). CODESRIA, Dakar, 297–342.

Wanjiru, M.W. (2018). Toponymy and the spatial politics of power on the urban landscape of nairobi, Kenya. Doctoral thesis, University of Tsukuba, Ibaraki [Online]. Available at: http://hdl.handle.net/2241/00152220 [Accessed 5 February 2020].

Woodman, P. (2015). The toponymy of absence. *Review of Historical Geography and Toponomastics*, 19–20, 7–16.

Worby, E. (1994). Maps, names, and ethnic games: The epistemology and iconography of colonial power in northwestern Zimbabwe. *Journal of South African Studies*, 20(3), 371–392.

Yermeche, O. (2018). Les dénominations de lieux en Algérie : entre toponymie coloniale, toponymie(s) officielle(s) et toponymie(s) spontanée(s). *Etudes et documents berbères*, 39–40(1), 387–400.

Zeleza, P.T. (1994). Review essay: Africa and the disintegration of paradigms. *Africa Development*, XIX(4), 179–193.

Zeleza, P.T. (2004). Africa, idea of. In *New Dictionary of the History of Ideas*, Horowitz, M.C. (ed.) [Online]. Available at: www.encyclopedia.com/history/dictionaries-thesauruses-pictures-and-press-releases/africa-idea [Accessed 12 August 2020].

# Conclusion

# Naming the Worlds

**Myriam HOUSSAY-HOLZSCHUCH[1] and Frédéric GIRAUT[2]**

[1] *Université Grenoble Alpes, CNRS, Sciences Po Grenoble, Pacte, France*

[2] *UNESCO Chair in Inclusive Toponymy "Naming the World", University of Geneva, Switzerland*

The rastaman thinks, draw me a map of what you see
then I will draw a map of what you never see
and guess me whose map will be bigger than whose?
Guess me whose map will tell the larger truth[1]?

The collection by the Jamaican poet Kei Miller, *The Cartographer Tries to Map a Way to Zion* (2014), is built around a face-to-face encounter between a cartographer and a rastaman. The former comes to measure, name, standardize and normalize. He speaks Oxford English, and his vocation is to map the world, as an enterprise of apparent objectification of physical realities and the colonial appropriation by the West of the Caribbean islands. The rastaman offers a geography that is at the same time anchored and spiritual, in permanent reconfiguration and re-creation, expressed in Jamaican English, recast by creolization and reworked by Rasta speech (*dread talk*) (Bailey 2019). This face-to-face encounter is at the same time a conversation, a ritual oral joust between representatives of two visions of the world, a confrontation between the colonized descendant of enslaved people and the one whose ancestors committed this crime, a Socratic dialogue inviting the cartographer to convert. Their exchanges, in a succession of single-speaker poems, are interspersed with other poems, including a series on place names, combining scholarly and vernacular etymologies, (multi-)

1. Kei Miller, *The Cartographer Tries to Map a Way to Zion*, 2014.

*The Politics of Place Naming*,
coordinated by Frédéric GIRAUT and Myriam HOUSSAY-HOLZSCHUCH. 

linguistic transformations, oral histories, banal experiences of wayside stops or terrible punishments under slavery. The exchange between them is complex, subtle and nuanced. The rastaman holds a thesis from the University of Glasgow, the same one which, while having an abolitionist position, earned millions in donations from the Atlantic slave trade, the same one which announced in August 2019 its willingness to pay reparations (Capella 2019). As for the cartographer, he sees new horizons opening up to him (Duff 2015), sketching out a possible role in restorative justice, redefining the map as accounting for spaces that are both experienced and imagined, a path towards something along the lines of revelation – Zion – rather than a presence in a place.

Miller's work will serve as a framework for this chapter, as it is so rich and speaks to questions of critical toponymy, those in which the field has become expert as well as those which remain to be explored and worked on. Miller's poetry also underlines the importance of thinking about questions of naming in multiple ways: multiple actors, multiple languages echoing one another – including in this text itself – multiple names, successive and simultaneous, for places with multiple histories and geographies.

The preceding chapters have shown this: critical toponymy, which deals with what naming places is, has established itself as an extremely fruitful field of research over three decades – a field that is both multidisciplinary (cultural and political geography, history, linguistics, semiology, town planning, cartography, sociology, anthropology and political science being the main protagonists) and deeply political. Today, this field is sufficiently solid to broaden its perspectives, even to proceed towards a profound theoretical renewal. All of the contributions to this book offer pathways towards this, which this concluding chapter attempts to weave together.

Indeed, the critical dimension of renewed studies of toponymy has until now been mainly expressed in the meticulous work carried out to repoliticize place names and the act of naming, by revealing the power relations at work there. Questions of political ecology (**Boillat**), relations of domination of gender and race, the (neo-/post-)colonial dimension where the act of naming is an integral part of processes of appropriation/dispossession (**Giraut**), the establishment of new political and spatial orders through toponymy (**Rose-Redwood et al.**) have been particularly well studied – as well as the increasing privatization of naming rights in the neoliberal era (**Vuolteenaho**). Conversely, the field of toponymy also appears as a possible site of reparations and restorative justice for past injustices (**Alderman**; **Noucher**), of political claims, in particular when metropolization or territorial reconfigurations cause the emergence of new places to name (**Destrem**; **Gauchon**; **Giraut and Houssay-Holzschuch**).

We should pursue this logic by integrating into the heart of critical toponymy the epistemological developments of Southern Theory (Robinson 2006; Connell 2007; Roy 2009). That is, by considering that the toponymic question depends not only on where it is situated in its empirical dimension (which acts of naming? by which actors? etc.), but also the anchoring, or better, the reterritorialization, of the toponymic question (its *situatedness*) causes its theoretical frameworks to fundamentally change. In other terms, and as **Ben Arrous and Bigon** invite us to do, it means taking with the utmost seriousness the theoretical changes necessitated by toponymic situations which are ontologically different from those of Western countries, post-socialist countries and settler colonies upon which critical toponymy was constituted as a field. African toponymic practices (**Wanjiru**) show different ways of doing or not doing between languages and places. In the same way as Lawhon (2020), Ben Arrous and Bigon show that this necessary attention to the epistemic dimension of place transforms the questioning: What is naming for the inhabitants? In which languages is it practiced? Which places and types of places are actually named? What is a place for whom, where and when? Why presuppose the unambiguity of the place/name relationship? Who names, when and with what success and audience? To take African toponymies seriously and, in general, toponymic questions outside the sites of production of critical toponymy as it has hitherto been established, is also to refuse to consider different situations in a teleological and Western-centric manner. These toponymic situations are not to be read as weaknesses, gaps to be filled, within the framework of a "development" orchestrated by those international bodies who are very active on the question, but an ontological diversity of a toponymic pluriverse (Escobar 2018), to be accepted as such in our theoretical constructions.

**ix. in which the cartographer travels lengths and breadths.**

Give him time and he will learn the strange
ways and names of this island [...]
will know the grunting path
to Hog Hole; the struggle required for Effort; the potholed
roads to Shambles, Rat Trap and Putogether Corner;
as well, the cartographer will know places named
after places – how this island spreads out as a palimpsest
of maps –: for here is Bethlehem; here is Tel Aviv; here
is Gaza; also Edinburgh; Aberdeen; Egypt; Cairo;
and here is Bengal; Mount Horeb; Albion; Alps;
they say – all of here is Babylon[2].

2. Ibid.

Therefore, we must first of all pay real, ontological attention to the places that we examine in the framework of critical toponymy. The latter has been traversed since its beginnings by the tension between the daily and lived experience of toponymy and the intense conflicts of renaming symbolic places – a country, a capital or metropolis, a major street, a mountain peak, between places of daily life and high places (Debarbieux 1993). This tension has not yet been dealt with directly: what place, what type of place, teaches us what? Ben Arrous and Bigon evoke sub-Saharan Africa as an epistemic area or site. Their argument should be extended to all sites, so that we might analyze how the specific geography of the study sites, beyond a spatial distribution favoring settler colonies (North America, South Africa, Australia and New Zealand), France, Israel and post-socialist countries (Giraut and Houssay-Holzschuch 2016), has configured our theoretical hypotheses and questions, and produced a number of blind spots. In a particularly interesting manner, the prominent place of Eastern Europe and the former USSR in the theoretical elaboration of critical toponymy foreshadows what Martin Müller (2020) calls for by proposing the notion of the *Global East*: taking these spaces, for which the term *post-socialist* is quite reductive, seriously as an epistemic site, like the thinkers of Southern Theory did for the Global South.

The hotspots and blind spots of the map of places studied by critical toponymy should also cause us to question ourselves on other levels: so far, we must admit a very strong urban bias in the existing literature. Obviously, practices of addressing in the West and beyond (Rose-Redwood, Tantner and Kim in this book), the metropolization and concentration of political debates on street names have constituted an urban toponymic corpus of remarkable variety and richness, which has allowed extremely fine analyses in terms of urban semiotics and understanding of the city as a text (*city-text*), cultural and political arenas of contestation of a gendered and racialized urban order, of performative space, to use the broad categories proposed by Rose-Redwood, Alderman and Azaryahu (2018). However, the peripheries of metropolization (rural and distant peri-urban spaces) remain understudied in critical toponymy, or are too often seen as simple places of extension of phenomena identified in the metropolitan core. Thus, the current extension of addressing in French rural municipalities, and more generally Western European ones (Karlander 2017), reveals power games that are structurally different from those in urban settings. Here, municipal actors who often define themselves as non-experts present the necessary naming processes as post-political – that is, as a technical necessity, of an administrative and bureaucratic nature – while in fact they entail debates and contradictions on what constitutes heritage and what deserves toponymic commemoration in a metropolitanized countryside inhabited more and more by neo-rural people or those who live as such. These debates and contradictions focus on remarkable landscape elements (remarkable for whom, and why?) or on local narratives (story-telling) and cultures.

If we continue to change scale, what theoretical blind spots are produced by the research focus on street names and therefore on ways and places alone? What would it mean to take equal consideration of other types of toponyms (the stathmonyms proposed by Destrem, the names of neighborhoods that are ultimately little studied despite their inventiveness, especially when informal, as noted by Wanjiru) or even rural microtoponymy (Boillat), the proper domain of a certain etymological or archaeological toponymy, but which has so far been less subject to theorization?

Even more, the issue of scale has not yet been addressed as such. From a theoretical point of view, what changes when we talk about country names, street names or plot names? The international dimension of toponymic conflicts is well known regarding the naming of gulfs, seas and straits, but their study has had little impact on our conceptual frameworks. This question of scale to be considered from a theoretical point of view is, in our view, an important element for deepening links between the field of critical toponymy and other fields sharing the same objects, such as international relations or urban studies (Rose-Redwood et al. 2018).

Finally, we know that toponyms circulate: "here is Bengal; Mount Horeb; Albion; Alps; they say – all of here is Babylon," says Miller. The necessary attention to places as epistemic sites must not be rigid, essentializing, occurring in a vacuum. The conditions of circulation of toponyms in their diversity, their scales of reference and the preferred languages of the circulating toponyms must in turn attract our attention. The influence of English and the languages of the Latin Mediterranean in the commodified and globalized toponymy, particularly visible in *malls* and gated communities, is a contemporary indicator and a powerful vector of this mobility, the diffusion and the relational character of neotoponymy. Equally important are the processes of reterritorialization of circulating toponymic references, which include strategies of boosterism or making claims.

**Place Name**

**Half Way Tree** – half-way between Greenwich and the brown
barracks of Stony Hill, 90 foot cotton under which soldiers
could ease boots from swollen feet, catch a lickle West Indian
breeze. Cue now a sound like feathers as letters retrieved from
back pockets, their contents already known by heart, are
unfolded and read again – words as fuel for the second portion
of journeying. And how that arboreal shade could make a script
more beautiful! So hail now the dearly departed half-way tree;

> the language tree; hail all that was read and written underneath; hail the loop of names carved as practice onto its wall-thick trunk; hail Mary O'Leary of Ulster and Allie Mac Dhughaill of Glasgow – the "dear loves" told to *come, come and smell the fruity rot of this place*; hail the tree first mentioned in 1696 which held its ground despite hurricanes, earthquakes, and even the British; hail tree which from its great height saw an island change. It fell in 1866 of good and natural causes. There rises now in its place, sepulchre-shaped, the Half Way Tree clock[3].

One of the decisive moves of critical toponymy was to displace analysis from the place name itself, as a finished product, to the process of naming as a whole, including names proposed but not adopted, debates, ephemeral namings, arguments and discourses of justification. This has precisely allowed the re-politicization if the field through bringing to light the power relations exercised there, whether they involve setting up or maintaining established orders, or indeed challenging them. The field today needs a second movement, of the same nature, which promises to be just as decisive.

Researchers in critical toponymy have been calling in recent years for a continued broadening of investigations, not only upstream of the name, but also downstream, and to take an interest in its reception. This must be envisaged with regard to the inhabitants of the named place, and also well beyond. Passers-by of all kinds (who do not reside in but pass through the place – to work, meet, have fun or as tourists), since places are also constituted by circulations, but also distant people for whom the practical and applied dimension of the name (getting one's bearings, making appointments, circulating) is not primary, also influence the way in which a name is perceived. This is a topic in which literature, especially since *Swann's Way*, has never ceased to be interested from the angle of sensory and existential perception (Arrivé 2013). In this context, we must not forget that the geoweb also widens the spheres of reception, particularly for vernacular toponyms (Choplin and Lozivit 2019). A place name bears stories and projections, as well as geographic imaginations. It evokes individual and collective experiences. In short, to use Lefebvre's terminology (1974), it is not only conceived in its process of naming by those with the official power to give names: it is perceived and experienced, individually and collectively. What a toponym does, what it produces, is still little known and more difficult to grasp from a methodological point of view. There is also no reason for this effect to be uniform across time, space and social relations, or even in biographical time.

3. Ibid.

Moreover, toponym users are not passive receivers: any reception implies a use, or rather several uses. Alternative, vernacular toponymies may be seen as responses or resistance to imposed toponymies. Similarly, diversions of form and meaning contribute to the development of corpuses (Kristol 2002). Do we choose to use the toponym, or do we use other systems of identification, for example, landmarks that are clearly visible in the physical or social landscape – the X café, the 63 bus stop, in front of the blue house? When several toponyms exist simultaneously for the same place, or have succeeded each other, which one do we choose to use, and why? Also, do we always use the same one? The use of a toponym is relational and situated; it depends on the context, what it will be used for, and between which interlocutors.

Critical toponymy would have much to gain here from a more complete inclusion of the kaleidoscope of emotions and memories, which condition reception, use and interpretation. Here, we are not referring only to the official interpretations of a toponym, to their political and public use – the referents may be biographical, familial or affective. In this context, critical toponymy needs a dialogue with the geography of emotions, and also with literary studies. In fact, there is no unambiguity of toponymy, but ambivalent and multiple meanings.

**What the mapmaker ought to know.**
On this island things fidget.
Even history.
The landscape does not sit
willingly
as if behind an easel
holding pose
waiting on
someone
to pencil
its lines, compose
its best features
or unruly contours.
Landmarks shift,
become unfixed
by earthquake
by landslide
by utter spite.
Whole places will slip
out from your grip[4].

4. Ibid.

The question of temporalities in the process of naming may also be enriched, insofar as toponymic studies tend to presuppose a certain stability in toponyms. A place name is largely conceived as being made to last, something which national toponymic administrations and international standards encourage and even demand. Thus, moments of toponymic change, of renaming, have particularly attracted the attention of critical toponymy, which displays great interest in them as revelatory of social and political factors. This practice of toponymic temporality as alternating long phases of stability and episodes of localized disturbances in time and space, however, tends to obscure other temporalities, which might also be revealing.

One of us (Houssay-Holzschuch 2018) has noted elsewhere that place names as such intrinsically mobilize diverse temporalities, thus reflecting what Doreen Massey (1992, 1995) called a *space–time*, where time and space are intrinsically linked. This is particularly noticeable with regard to odonyms or street names:

– Due to their commemorative character, they inject the past into the present.

– Furthermore, when a street has changed names several times, several pasts are invoked when it is named: the period commemorated by the odonym, but also the period of its official use.

– Toponymic changes also have specific temporalities – conquest, change of political regime, post-situations (colonial, socialist, apartheid, conflict), etc. (Giraut and Houssay-Holzschuch 2016).

– The uses and interpretations of street names involve questions of memory, collective and individual, of urban changes, as well as of nostalgia.

A toponym's regime of temporality is therefore complex.

Moreover, as Duncan and Young remind us (Rose-Redwood et al. 2018), focusing on toponymic changes should not exclude analysis of what they call "'left-over' toponymic landscapes", the continuities of the toponomascape, the names that do not change and their characteristics. In a pioneering study, Horsman (2006) has in fact shown that the peaks of the Pamir, subject to regular renaming because of Russian political changes, were not all renamed in the same way. The highest, the most visible, the most central in the massif underwent the greatest number of toponymic changes, while more peripheral peaks, with names that were just as marked from an ideological point of view, remained under the radar of toponymic cleansing. On the literary side, Milan Kundera constructs his entire novel *The Festival of Insignificance* around the absurd retention of the name of Kaliningrad for the former Königsberg, birthplace of Immanuel Kant. This retention, for the novelist, can only be explained by the insignificance of its referent – Mikhail Kalinin – already the most despised of Stalin's comrades at the time when he renamed the city after him.

Finally, temporary regimes of toponymic temporalities appear more frequently, and demonstrate the value of taking an interest in this dimension. The commodification of naming rights, branding by toponym, opens the door to significant toponymic turnovers, whether because the naming rights expire, or because the brand that acquired them wishes to proceed with a rebranding, as advised by marketing agencies. In a very different field, that of contestation and demands for toponymic and social justice, toponymic "coups" or performances designed to be ephemeral, pasting alternative names on top of contested names, have shown their strategic and political interest. The diversity of regimes of toponymic temporalities, which also include regimes of simultaneous toponymic temporalities (several names coexisting for a single place), should thus be considered as such and analyzed from a theoretical point of view.

**In which the rastaman disagrees.**
The mapmaker's work is to make visible
all them things that shoulda never exist in the first place
like the conquest of pirates, like borders,
like the viral spread of governments[5].

Finally, critical toponymy would benefit from gauging the extent of state-centrism in the manner of approaching the process of naming. This question is posed in a broader way, to critical geography, for example, which Barrera de la Torre (2021) engages to "stretch geography's state-centric epistemic constraints" for a "state-decentering epistemological move in the discipline". In fact, as Barrera and Ince remind us (2021), "the state and geography have been central to each other's institutionalization, and the state has shaped in many ways the epistemic standards that continue to dictate how geographical knowledge is produced". For formal naming processes, the state is central, including for mundane places and the everyday landscape (Billig 1995). Yet the way in which it officially names, or validates toponyms proposed by civil society, at the junction of political and bureaucratic authorities and in the multiplicity of its identities (Scott 1998; Möser and Tillous 2020), deserves more critical attention and, no doubt, more interdisciplinarity with the sociology of organizations: what, precisely and concretely, is the naming state? What does it do and who does it? Similarly, the strictly international dimension, in the sense of interactions between established states, international toponymic institutions and bureaucracies, has not been critically considered. Despite a number of efforts, the decision-making process of the actors remains to be elucidated to discover who officially gives what type of name, when, why and to what type of place. Proceeding by institution or pressure group, as has

5. Ibid.

most often been done, remains a rather coarse form of granulometry, which must be refined in order to identify precise, even individual actors, including in the name-giving bureaucracy. Scenes of contestation appear here as fantastic terrains: judicial intervention; participation in the selection or validation processes; contradictory use of cartography; political and artistic performances.

Critical toponymy is very interested in changes in the role of the State in a neoliberal regime, analyzing devolution to reconfigured local powers (a new administrative geography whose toponymic power has been particularly studied in the Francophone context; see Giraut and Houssay-Holzschuch 2008), renting the right to name, official validations of private naming in the context of real estate projects, etc. This volume bears witness to this. However, we also know that this role of the neoliberal state is one of reconfiguration and not of withdrawal. What has become today of its traditional toponymic prerogatives and their real functioning? What are the contours of the naming state, to be conceptualized, as proposed by Davina Cooper (2020), in a situated, contingent and anti-essentialist way?

Finally, taking a step back from disciplinary statocentrism would make it possible to fully integrate unofficial, bottom-up toponymies, as well as various state-produced forms, into theoretical reflection — and not in the teleological mode of a lack. In short, to take up the title of Möser and Tillous (2020), to work explicitly on a critical toponymy "with, without or against" the state.

So every night while the mapmaker expands
on his network of secret roads and slaving roads,
marooning roads and backbush roads,
what he has really concerned himself with is Zion –
a question has wedged itself between his learning
and awakening: how does one map a place
that is not quite a place? How does one draw
towards the heart[6]?

## References

Arrivé, M. (2013). Toponymie et littérature. *Approches interdisciplinaires de la lecture.* Presses universitaires de Reims, Reims.

Bailey, C. (2019). Voicing the reclamation in Kei Miller's *The Cartographer Tries to Map a Way to Zion. Sx salon,* 32.

6. Ibid.

Barrera de la Torre, G.N. (2021). Introduction: Anarchist geographies and epistemologies of the state. *ACME: An International Journal for Critical Geographies*, 20, 142–150.

Barrera de la Torre, G.N. and Ince, A. (2021). Society despite the State: An experiment in "Counterfactual Statism". *ACME: An International Journal for Critical Geographies*, 20, 214–221.

Billig, M. (1995). *Banal Nationalism*. Sage, London

Capella, H. (2019). Glasgow University's "bold" move to pay back slave trade profits. *BBC News*, 23 August.

Choplin, A. and Lozivit, M. (2019). Mettre un quartier sur la carte : cartographie participative et innovation numérique à Cotonou (Bénin). *Cybergeo: European Journal of Geography*. DOI: https://doi.org/10.4000/cybergeo.32152.

Connell, R. (2007). *Southern Theory. The Global Dynamics of Knowledge in Social Science*. Polity, Cambridge.

Cooper, D. (2020). Des identités multiples. La sexualité et l'État en lutte. In *Avec, sans ou contre. Critiques queer/féministes de l'État*, Möser, C. and Tillous, M. (eds). Éditions iXe, Donnemarie-Dontilly.

Debarbieux, B. (1993). Du haut lieu en général et du mont Blanc en particulier. *L'Espace géographique*, 22(1), 5–13.

Duff, V. (2015). Mapping Kei Miller's Zion. *The Critical Flame. A Journal of Literature & Culture*, 34 [Online]. Available at: http://criticalflame.org/mapping-a-way-to-kei-millers-zion/.

Escobar, A. (2018). *Designs for the Pluriverse: Radical Interdependence, Autonomy, and the Making of Worlds*. Duke University Press, Durham.

Giraut, F. and Houssay-Holzschuch, M. (2008). Néotoponymie : formes et enjeux de la dénomination des territoires émergents. *L'Espace politique*, 5–12.

Giraut, F. and Houssay-Holzschuch, M. (2016). Place naming as dispositif: Toward a theoretical framework. *Geopolitics*, 21(1), 1–21.

Horsman, S. (2006). The politics of toponyms in the Pamir mountains. *Area*, 38, 279–291.

Houssay-Holzschuch, M. (2018). The political life of urban streetscapes: Naming, politics, and place–book review. *The AAG Review of Books*, 6, 257–259.

Karlander, D. (2017). Roads to regimentation: Place, authenticity and the metapragmatics of naming. *Language & Communication*, 53, 11–24.

Kristol, A.M. (2002). Motivation et remotivation des noms de lieux : réflexions sur la nature linguistique du nom propre. *Rives nord-méditerranéennes*, 11, 105–120.

Lawhon, M. (2020). *Making Urban Theory: Learning and Unlearning through Southern Cities*. Routledge, London.

Lefebvre, H. (1974). *La production de l'espace*. Anthopos, Paris.

Massey, D. (1992). Politics and space/time. *New Left Review*, i, 65–84.

Massey, D. (1995). Places and their pasts. *History Workshop Journal*, 39, 182–192.

Miller, K. (2014). *The Cartographer Tries to Map a Way to Zion*. Carcanet, Manchester.

Möser, C. and Tillous, M. (eds) (2020). *Avec, sans ou contre. Critiques queer/féministes de l'État*. Éditions iXe, Donnemarie-Dontilly.

Müller, M. (2020). In search of the global east: Thinking between north and south. *Geopolitics*, 25, 734–755.

Robinson, J. (2006). *Ordinary Cities: Between Modernity and Development*. Routledge, New York.

Rose-Redwood, R., Alderman, D., Azaryahu, M. (eds) (2018). *The Political Life of Urban Streetscapes: Naming, Politics, and Place*. Routledge, New York.

Roy, A. (2009). The 21st-century metropolis: New geographies of theory. *Regional Studies*, 43, 819–830.

Scott, J.C. (1998). *Seeing Like a State: How Certain Schemes to Improve the Human Condition Have Failed*. Yale University Press, New Haven.

# List of Authors

Derek H. ALDERMAN
Department of Geography & Sustainability
University of Tennessee
Knoxville
USA

Michel BEN ARROUS
Gaston Berger University
Saint-Louis
Senegal

Liora BIGON
Holon Institute of Technology
Israel

Sébastien BOILLAT
University of Bern
and
Bern University of Applied Sciences
and
University of Neuchâtel
Switzerland

Lucas DESTREM
Independant Geographer
Limoges
France

Christophe GAUCHON
Edytem
Université Savoie-Mont-Blanc
Chambéry
France

Frédéric GIRAUT
UNESCO Chair in Inclusive Toponymy
"Naming the World"
University of Geneva
Switzerland

Myriam HOUSSAY-HOLZSCHUCH
Université Grenoble Alpes
CNRS, Sciences Po Grenoble, Pacte
France

Sun-Bae KIM
Chungju Girls' High School
South Korea

Matthieu NOUCHER
CNRS Passages
Bordeaux Montaigne University
France

Reuben ROSE-REDWOOD
Department of Geography
University of Victoria
Canada

Anton TANTNER
Institut für Geschichte
Universität Wien
Vienna
Austria

Jani VUOLTEENAHO
Faculty of Social Sciences
University of Helsinki
Finland

Melissa WANJIRU-MWITA
Technical University of Kenya
Nairobi
Kenya

# Index

## L, M



## N, O



## P, R



## S, T

## V, W

Printed and bound by CPI Group (UK) Ltd, Croydon, CR0 4YY
13/09/2026
14960999-0002